Practical Astronomy

Lectures on Time, Place, and Space

David H. DeVorkin

Illustrations by **Steve Postman**

Published for the **National Air and Space Museum**
by the **Smithsonian Institution Press**
Washington, D.C., and London, 1986

Library of Congress Cataloging-in-Publica-
tion Data

DeVorkin, David H., 1944–
Practical astronomy.

Bibliography: p.
Includes index.
Supt. of Docs. no.: SI 9.2:T48
1. Astronomy, Spherical and practical. I.
National Air and Space Museum. II. Title.
QB145.D48   1986   522   86-42648
ISBN 0-87474-359-1 (alk. paper)

Contents

Introduction

This book of lecture notes on practical astronomy has been prepared to enhance popular understanding of the astronomical basis for the reckoning of time, the seasons, and navigation. Other topics amenable to planetarium lectures included here are the visible structure of the universe, determining the scale of the visible universe, and observational clues to the dynamic behavior of the solar system.

Designed as a supplement to a set of eight lectures for the public and for school groups visiting the Albert Einstein Planetarium at the National Air and Space Museum, the book should also work as a self-study guide or as a mini-course at the high-school or introductory college level.

The topics are within the traditional framework of classical astronomy, derived from studies of the visible sky that have been carried on by many cultures for thousands of years. While they represent tradition, they do not represent the modern activities of most members of the world's community of astronomers. Today, very little professional attention is paid to the study of the visible sky. Most observing astronomers peer deep into space with sophisticated detectors attached to huge telescopes or orbiting spacecraft to determine the physical nature of the objects found there. Photoelectric detectors, radio telescopes, spectrographs and other exotic devices have all but replaced the eye at the end of the telescope. And telescopes themselves are controlled by computer-driven automatic systems that further remove the observing astronomer from the visible sky.

Most astronomers now spend far more time analyzing data, developing theoretical models, or even writing grant proposals than they do contemplating the visible sky and the many apparent motions of the sun, moon, and planets that have fascinated mankind since prehistory.

The majority of modern college-level textbooks on astronomy, even if designed for those not contemplating a career in science, reflect this modern character of the discipline. Thirty years ago, textbooks commonly included a substantial discussion of visible sky astronomy, but in recent years, this discussion has dwindled to the point where it has become a brief preamble, not to be taken to heart. Modern developments in astronomy are highly provocative and fascinating, and well it might be that more attention is devoted to the weird and wonderful things we now know to exist in the universe. On the other hand, this modern emphasis has divorced many people from the very personal contact they have had with astronomy—through the visible sky. These notes, and the lectures they support, are intended to provide access once again to this very important point of contact.

Clearly, then, this is not an introduction to modern astronomy, as practiced by most astronomers today, but the subject matter forms much of the basis for modern astronomy, especially the latter lectures on the dimensions of space and the dynamics of the solar system. These lectures, while dealing with historical topics, have been presented in an ahistorical fashion, for the sake of getting across in a simple manner the essence of the processes involved. A bibliography provides access to more complete scholarly treatments of the historical subject matter.

The lecture notes themselves were developed from years of teaching the subject in planetariums at UCLA, San Diego State University, Central Connecticut State University, and at the National Air and Space Museum. The artwork has been deftly executed by Steve Postman, of Washington, D.C. The Smithsonian Office of Elementary and Secondary Education and the National Air and Space Museum's Office of Education kindly supported the publication of this book.

David H. DeVorkin
Chairman, Department of Space Science and Exploration
National Air and Space Museum

1 Orientation of the Earth in Space

As you look up into space at night, the stars, the moon, and the planets seem to be placed upon an inverted bowl that touches the earth in a wide circle centered upon your place of observation. This circle is called the **horizon.**

After an hour or two, careful examination of the sky will reveal that the stars are all in motion. They seem to be rising over one part of the horizon, the **east,** and setting behind another, the **west.** There is also a place where stars achieve a maximum distance above the horizon, the **south,** and a place where the stars seem to be moving very little or not at all, the **north.** These four places are called the **cardinal points of the compass:** east, west, south, and north (see figure 1).

The apparent westerly motion of the stars on the **celestial sphere,** or the inverted bowl that we call the sky, is called **diurnal motion,** and is really an illusion caused by the rotation of the earth upon its axis. It is easy enough for us to say this, and easy enough for you to remember this, but have any of us actually experienced this motion? Can we feel this motion of rotation? The short answer is no. This, in fact, was one of the greatest difficulties that had to be overcome before we were able to think of the earth in motion, not only in rotation about its axis (producing the day) but in revolution about the sun (producing the year). Not detecting these motions, our ancestors believed that all motion seen in the heavens was actual motion, and that the earth was stationary in the center of the cosmos.

The Earth's Rotation

Diurnal motion can be explained equally well by a stationary earth and a moving celestial sphere, or by a moving earth and a stationary celestial sphere. To demonstrate this in a planetarium, we project ourselves into the earth and look out at the earth's surface from a

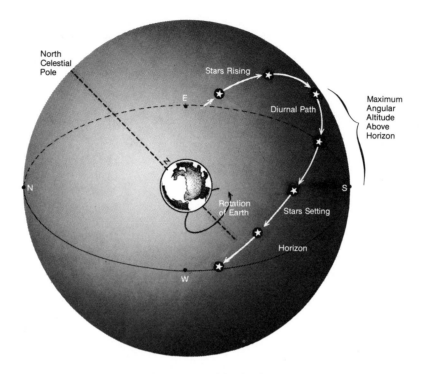

Figure 1 The celestial sphere and its basic components.

point close to its very center. Above us now would be the continents, seen reversed east to west, of course, because we are looking at them from the inside. Beyond the continents we can still see the stars.

First, notice where the earth's **equator** is. It is quite visible as a broad line. If we were to allow apparent movement of the stars by moving the celestial sphere, we would first notice that the stars move parallel to the earth's equator. Here, then, we see the stars moving over the earth's surface from east to west. Any one star will be above eastern points on the earth's surface before it is above western points. Specifically, we will see any star pass over us in Washington, D.C., well before it is seen passing over Denver, Colorado, or Los Angeles, California.

Now we stop the celestial sphere and allow the earth to rotate upon its axis. First, notice that the continents are in motion, and that they move with the earth's equator. Notice, too, that the earth rotates from west to east. The stars are now stationary. Try to imagine what this looks like from the moving surface of the earth. Pick out any bright star. See how it apparently travels, or, how the earth travels underneath it. Eastern points on the earth will be under it first, and then western points. This is really the same effect that we had before when the celestial sphere was in motion and the earth was stationary.

Let us now project ourselves back onto the earth's surface and watch what happens. As we jump onto the moving earth, the stars immediately begin to move "backward"—a reflection of our motion. So which is the true cause of the diurnal motions of the stars? From this, we cannot tell. The best we can do is offer a physical demonstration first performed by the Frenchman Leon Foucault.

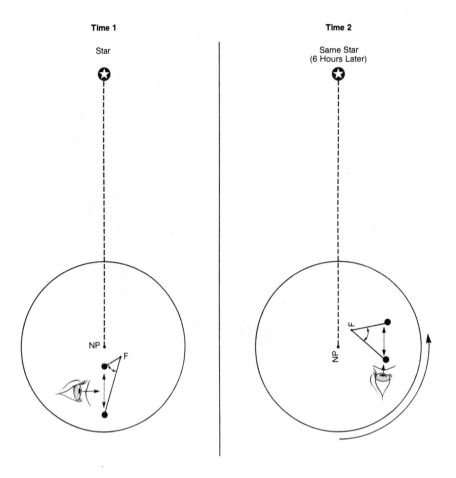

Figure 2 Looking down on the North Pole of the earth, a pendulum set in motion at an intermediate latitude on the earth's surface will always swing in the same direction with respect to the stars, but will apparently shift in direction when seen by an observer on the earth.

The Foucault Pendulum

To appreciate this demonstration fully, we need Isaac Newton's laws of motion, which will be discussed later in the text. For our demonstration now, we only need point out that as a consequence of one of Newton's laws, a pendulum, once set swinging in one direction, will always swing in that direction.

By "direction" here we do not mean "north, south, east, or west" but a course with respect to some larger, more fundamental, frame, such as one defined by the stars. The direction of the pendulum's swing will, therefore, be independent of the earth's motions, specifically its rotation. If a pendulum is set in motion, and remains swinging for some time, its independence from the motion of the earth will cause its direction of swing to appear to be moving to anyone looking at it who is standing on the moving earth. The 19th-century French physicist Leon Foucault [Foo-coe] was the first to exploit this.

As figure 2 illustrates, the moving earth rotates "underneath" the pendulum and carries

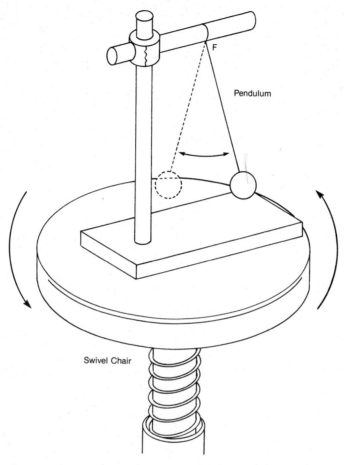

Figure 3 Demonstration of the independence of pendulum swing from the motion of the earth, as represented by a spinning swivel chair.

the observer from a position perpendicular to the direction of the pendulum's swing to a position in line with the direction of the swing. All the while, of course, the direction of swing of the pendulum has remained fixed with respect to the distant stars, in accordance with Newton's first law of motion.

A simple demonstration of the Foucault pendulum requires a swivel chair without a back, a few C-clamps, and some standard metal beaker supports found in any physics or chemistry lab. The setup only requires that a small pendulum can be put in motion, swinging at a point directly above the axis of rotation of the swivel chair, which represents the rotating earth. Set the pendulum swinging (see figure 3) and then rotate the chair a bit (about 45 to 90 degrees). The original direction of the pendulum's swing should not be perceptibly altered by rotating the swivel chair. In practice, of course, the pendulum is less than perfect, so the plane of swing will shift a bit, but this is due to differential torsion on the pendulum's string where it meets the supporting arm. If a "perfect" system without torsion were available, no rotation would occur.

Pleiades

Hyades

Procyon

Aldebaran

Betelgeuse

ORION

TAURUS

Sword

Rigel

Sirius

SOUTHERN HORIZON

Figure 4 Orion and Taurus the Bull, as seen from the northern mid-latitudes on a winter evening.

8

The Visible Sky

Now that we have some idea of why we see the stars move in their nightly courses, and why we really cannot sense our own motions, we will return once again to the night sky.

Let us imagine we are on the surface of the earth, and have the stars above our heads. We will point out a few celestial landmarks, a few stars and constellations that will help us become better oriented to the night sky.

The names of stars come to us from antiquity. Like constellations, they are products of the imagination and reflect characters out of the rich mythological tradition of our cultural ancestors. Excellent references in our bibliography for the roots of these star and constellation names include *Star Names,* by R. H. Allen, and works on mythology by Thomas Bulfinch and Edith Hamilton.

Among the most famous constellations in our winter sky is the mighty warrior Orion. Four stars mark his shoulders and feet, and three mark his belt. He has a sword and shield, and is fighting a bull (figure 4). The stars that make up this constellation are in reality quite different from one another. Most are not associated with one another in space and are of different ages and different masses. Betelgeuse (which translated loosely from Arabic means "arm pit") is a red giant, vastly larger than the sun. Rigel (or "foot") is an extremely hot blue star, as are the stars in the belt.

In the sword of Orion is a faint hazy patch that isn't a star at all. Under magnification in binoculars or a modern telescope, it looks like a vast gaseous cloud and is believed to be the spawning ground for stars. Modern astronomers are fascinated with this Great Nebula, as it is called, for it is through studying regions of space like this that we are able to understand how stars are born, how they live, and how they eventually die.

Constellations close to Orion are Taurus the Bull, Auriga the Charioteer, Gemini the Twins, the Pleiades, and, of course, Orion's faithful hunting dogs, Canis Major and Canis Minor. The Pleiades actually consist of a physical cluster of stars called a **galactic cluster;** these stars are much younger than the sun and the remnants of their birth are visible in long exposure photographs taken with telescopes. In Taurus, another cluster called the Hyades is faintly visible. Most stars in Taurus' head make up the Hyades. This is one of the closest galactic clusters to the sun.

In the winter months these constellations rise close to the eastern point of the horizon in the early evening, travel across the sky, are easily visible above the south close to midnight, and set in the west near dawn. We will later see that these constellations are best visible only in the winter months. If we look farther north, however, other constellations and stars will be found that do not behave this way.

Circumpolar Constellations

Watching the northern sky as the earth rotates, we can see that the stars in the north circle one particular point in space very close to a star called Polaris, which is the last star in the handle of the Little Dipper. From mid-northern latitudes, the stars of the Little Dipper never rise or set.

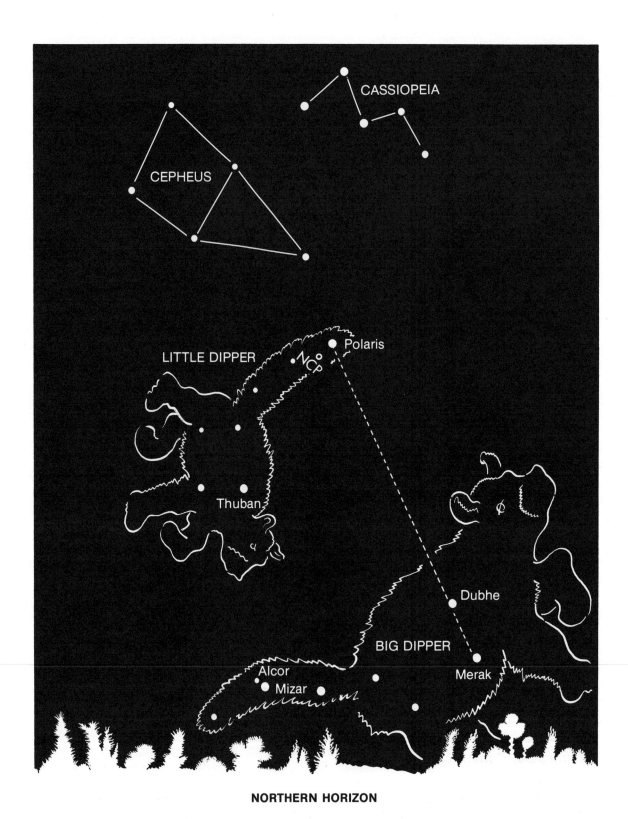

Figure 5 The circumpolar sky as seen on an early winter evening.

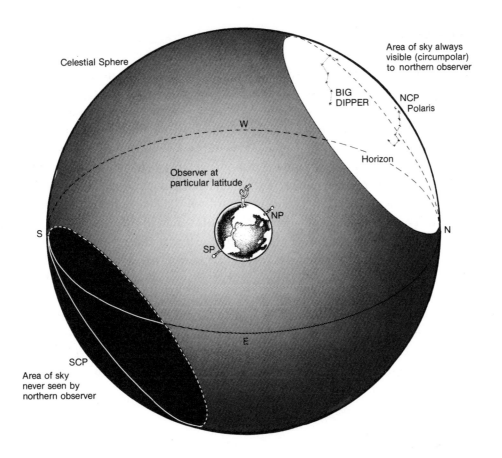

Figure 6 The shaded areas represent parts of the sky always visible (around the NCP) or never visible (around the South Celestial Pole, or SCP), as seen by an observer in a mid-northern latitude.

They apparently circle the **North Celestial Pole,** or **NCP** and are therefore called **circumpolar stars** (figure 5). The Little Dipper, for example, is a **circumpolar constellation.**

Several other constellations are close enough to the NCP to be circumpolar. The most prominent is the Big Dipper, which is really only part of a larger constellation called Ursa Major, or the Great Bear. This constellation can be used to find the NCP. The two stars in the front of the bowl of the Dipper point directly to Polaris, moving in the direction that the Dipper bowl opens. The NCP is that part of the celestial sphere directly above the North Pole of the earth (see figure 6).

The constellations Little Dipper, Big Dipper, Cassiopeia, Draco, and Cepheus all circle the NCP and never rise or set in mid-northern latitudes. For more southerly latitudes, fewer and fewer of these constellations are circumpolar.

Spring Constellations

In the spring months (late March through June), the dominant configuration of evening constellations includes the Big Dipper, directly overhead, or close to the **zenith,** and Leo, Bootes, Virgo, Corona Borealis, and Hercules (see figure 7). The winter constellations Auriga, Gemini, and Canis Minor are all close to the western horizon and therefore are setting in the evening. Orion sets just after the sun in mid-May, but in late March, sets close to midnight.

Starting with the Big Dipper in figure 7, follow the arc of the Dipper handle to the first bright star. This will be the reddish yellow star called Arcturus ("follow the arc to Arcturus"), which is the brightest star in Bootes. Continue to follow the arc, and the next bright star (a beautiful blue) will be Spica, in Virgo. Using the pointer stars in the bowl of the Big Dipper, move in the direction of the bottom of the bowl, and the first major group you come to will be Leo; the brightest star, at the base of the reversed question-mark pattern, is Regulus.

Due east of the Big Dipper is Hercules, looking like a very faint Orion. It is halfway between the bright stars Arcturus and Vega, which at this time of year and hour in the evening are low in the eastern sky. In the left side of Hercules (stage left) is a magnificent telescopic object called a **globular cluster.** It might be just barely visible as a faint fuzzy patch in a very dark moonless sky, if your eyes are very well adapted to the dark. This vast spherical cluster of stars can be detected with binoculars reasonably well, but a large telescope is required to show that it is not a nebula but is in fact composed of stars.

Summer Constellations

In the early evening during the summer months (June through September), high in the eastern sky, are three bright stars that make up what some like to call the "Summer Triangle" (see figure 8). Actually, each star is in a separate constellation: Vega, the brightest of the three stars, is in Lyra the Harp; Deneb, in Cygnus the Swan, sitting right in the plane of the **Milky Way** (the visible portion of our **galaxy**); and Altair is in the constellation of Aquila the Eagle.

Cygnus is the most impressive constellation of the three. To some it is a swan, flying south along the Milky Way (anticipating the fall months?). In fact, Deneb means "tail of the swan." To most people today, Cygnus is called the "Northern Cross," for that is what it really looks like. The main staff starts with Deneb and moves through three stars in a southerly direction to Alberio, the base of the main staff, or head of the swan. In a small telescope, Alberio is a beautiful optical double star—the fainter component is vivid blue, and the other is bright gold.

In Lyra, which is a small harp, there is really nothing to see unless you have a telescope. A properly equipped telescope of about four-inch aperture or larger should be able to pick up a small planetary nebula, called the Ring Nebula, between Beta and Gamma Lyrae. This object is a star at the end of its life. In the process of star death, it is releasing atmospheric gases in a vast and expanding envelope, which looks like a ring in projection. Another interesting object in Lyra is the faint star Epsilon Lyrae, which is in fact a "double-double" star system consisting of four stars in two sets of two.

Two of the most beautiful constellations in the entire sky are just barely visible above the southern horizon during the summer months. They are best seen in July around 10 p.m., eastern daylight savings time. Scorpius, low in the sky, is a large scorpion, starting as a hammerlike head and claws, with a long curving body passing through the bright red star Antares and curving up into a little stinger. Just to the east of Scorpius is Sagittarius the Archer. To our ancestors it was a centaurlike creature, half man, half beast, complete with a bow and arrow. Today, people like to see it as a teapot, with a small handle, top, body, and spout. In fact, the steam coming from the spout can be seen too: it is the bright southern Milky Way!

Lines in the Sky

The apparent position of the North Celestial Pole (NCP) in our sky is determined by our position on earth, specifically, by our **latitude.** Washington, D.C., is approximately 38 degrees north, which means that we are a little less than half way to the North Pole, as seen along a circle starting from the equator, traveling through Washington and ending at the North Pole.

Such a circle—really an imaginary line on the surface of the earth—is called a **meridian of longitude.** This meridian can be projected beyond the earth onto the celestial sphere. It then becomes a **celestial meridian,** and for each longitude on the earth, one of these celestial counterparts exists. Each one of us has one. We have been carrying it around since birth. Our personal meridians are called the **observer's meridian.** We can project one for all of us onto the celestial sphere, starting at the south point, traveling vertically until it reaches the zenith, continuing through the NCP, and ending at the north point of the horizon (see figure 9).

Positions on this meridian are marked off in degrees; they represent angular distances from the north and south points on the horizon. These are **altitudes,** which extend to 90 degrees at the zenith. In Washington, D.C., the altitude of the NCP is approximately 38 degrees above the north point. This is, of course, equal to our latitude. We might therefore conclude that the altitude of the NCP reveals our latitude on earth. In fact it does. But to understand this relationship we must look closer and examine the general case, as in figure 9.

Figure 9 depicts the basic celestial sphere. The observer is at the center, with the horizon plane stretching out in all directions to the horizon. The observer's meridian is in the plane of the page marking the boundary between the eastern sky and the western sky. The **celestial equator** is also seen, starting at the east point, rising to a maximum altitude above the south point, and continuing on to the western horizon. It continues on below the horizon as well, making a complete circle.

The altitude of the NCP can be measured in two ways. First, the angular altitude can be determined by the length of a celestial arc from the north point of the horizon to the NCP. Second, the angle produced by two lines of sight from the observer, one to the north point and the other to the NCP is also the altitude (angle A). The equivalence of these two angles is important in our study. Other examples will be seen later.

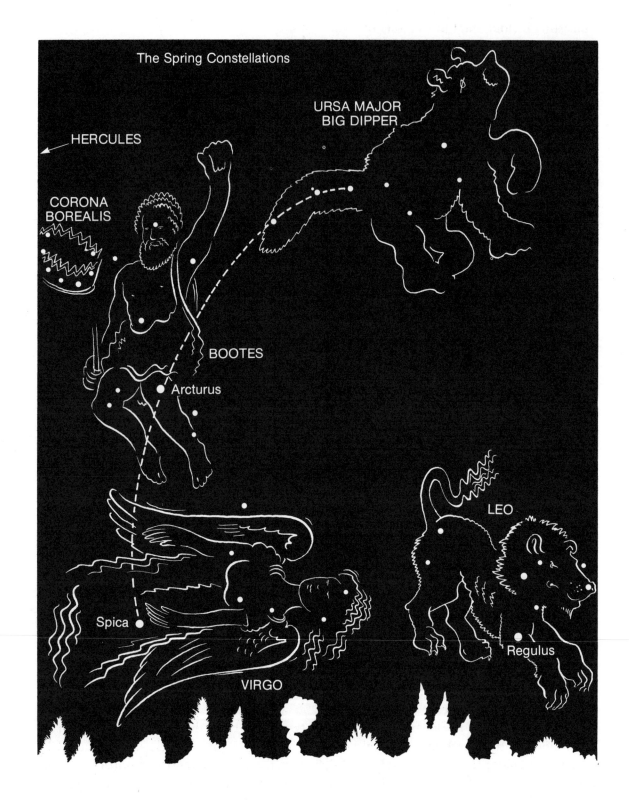

The Spring Constellations

HERCULES

URSA MAJOR
BIG DIPPER

CORONA
BOREALIS

BOOTES

Arcturus

LEO

Spica

Regulus

VIRGO

Figure 7 The six spring constellations.

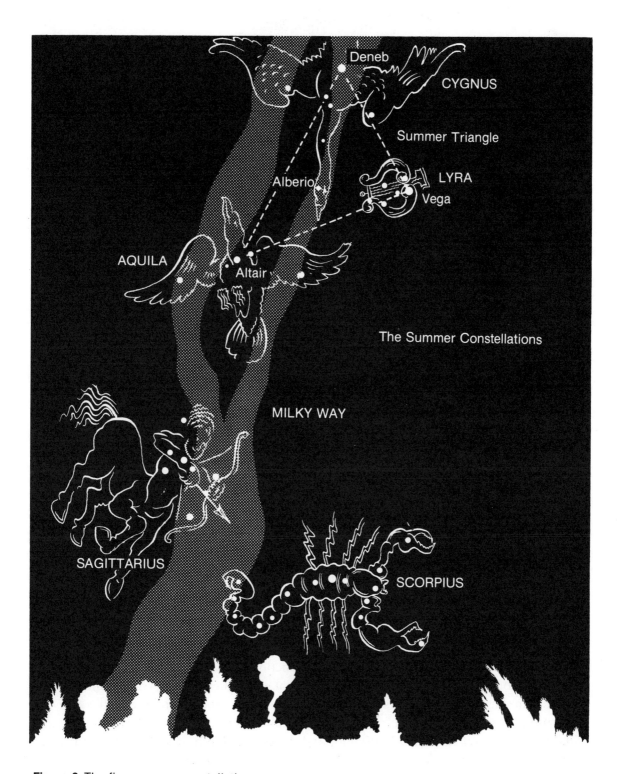

Figure 8 The five summer constellations.

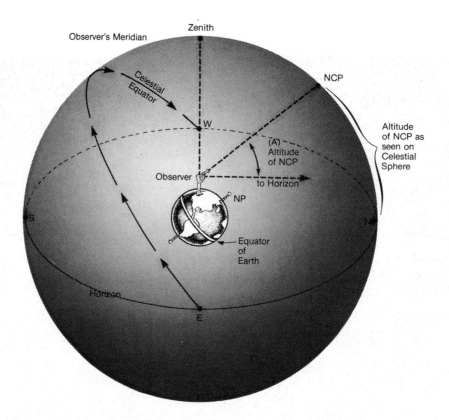

Figure 9 The fundamental orientation of the earth in space is defined by the direction its axis of rotation points. The point of intersection of an extension of this axis of rotation and the celestial sphere is the North Celestial Pole (or NCP). A circle joining the NCP, the north point of the horizon, the observer's zenith, and the south point of the horizon is called the observer's meridian. It is a great circle inscribed onto the celestial sphere, but not fixed to the sphere.

The Celestial Sphere

The celestial sphere is an idea that evolved thousands of years ago in an attempt to comprehend the real sky. To our distant ancestors, it was a crystalline globe rotating around the earth. Small fissures and cracks in the globe (so some romantic tales go) allowed light from heaven to stream through. These were the stars. In some ancient cultures, the sky was a vault, or canopy, held up by vast mountain ranges, and the stars were deities suspended from the canopy. To all ancient minds, the firmament was a strange and wonderful place, the place of gods.

In modern thought, the celestial sphere is a functional device with very real conceptual value, but also real limitations. It is two-dimensional, while the real sky is three-dimensional. Thus it is an illusion, but it allows us to map the sky and to understand the earth's celestial orientation. To speak of the "size" of this celestial sphere is meaningless. The moon is a mere 250,000 miles away; the sun, 93,000,000; and the stars are at hundreds of thousands and millions of times greater distances. Yet they all appear to be on the celestial sphere and to travel with it, or on it, during the course of day and night.

To examine the earth's orientation in space, we must assume that the celestial sphere is of infinite size. This is, of course, impossible to visualize, but the following exercise will help. If we imagine a series of "celestial spheres," each of increasing size with respect to the earth, we can see how points on these spheres differ in position as seen from different parts of the earth (see figure 10). As the celestial sphere is made larger and larger, any point on its "surface" is seen in almost the same direction from two points on earth labeled A and B. The larger the sphere, the smaller the discrepancy in position. If the sphere were to be made infinite, the angle would be zero and the lines AS and BS would be *parallel*.

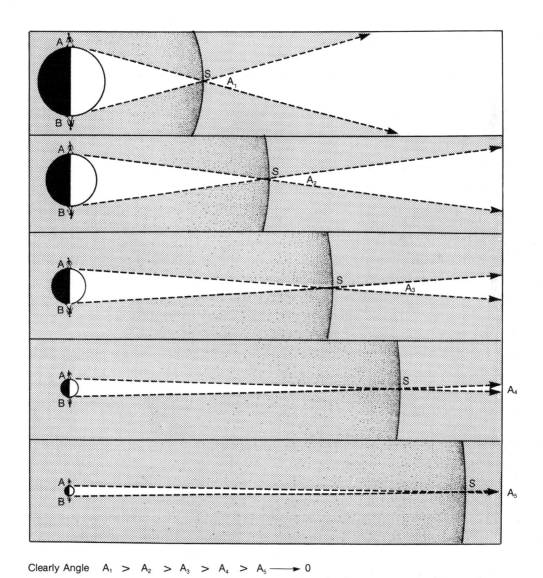

Clearly Angle $A_1 > A_2 > A_3 > A_4 > A_5 \longrightarrow 0$

Figure 10 "Celestial spheres" of different sizes, extending to infinite size. As the spheres get larger, lines of sight from different points on the earth to one point on the sphere become parallel. Any point on an infinite celestial sphere will appear in the same direction from different points on the surface of the earth.

This is the primary rule that we will use in our study of the celestial sphere: that *the direction to any point on the celestial sphere is the same as seen from any point on the earth*, or, that *all lines of sight drawn from any points on earth to any one point on the celestial sphere are parallel.*

Now, to the problem at hand—demonstrating that the altitude of the NCP is equal to our latitude. First we must draw the celestial sphere in profile projected with the observer's meridian in the plane of the page. The earth must be drawn of sensible size, but it *must* be understood that the celestial sphere is infinitely larger than the earth.

Figure 11 illustrates the situation. First, the observer is placed on the earth ("observer" on diagram) and directions are drawn to the NCP, zenith, and the celestial equator. Directions to the cardinal points (North, South, East and West) are also given. These directions from the center of the earth are also given and they are parallel to the directions from the observer's position. Angle L represents the latitude of the observer, or the observer's angular distance from the earth's equator. Since this angle makes up part of a 90-degree angle between the equator and the axis of rotation of the earth (in the north direction), the remainder of the right angle is 90 degrees – L.

This second angle is in a triangle formed by the center of the earth, the observer, and the intersection of the north horizon and the line to the NCP from the earth's center (point P). This is a right-angle triangle, whose sum of angles must be 180 degrees. Since 90 – L is one angle, and the angle at the observer is 90 degrees, the third angle must be L degrees. Now, the line from the observer to the NCP makes an angle A with respect to the line from the observer to the north point of the horizon. Since the line (observer to NCP) is parallel to the direction to the NCP from the earth's center, we have a geometric situation where two parallel lines are crossed by a third line (OP), producing a set of equal **alternate interior** angles. Thus the angle between the north horizon and the line to the NCP from the earth's center is also angle A. But since angle A is a vertical angle with respect to angle L in the triangle mentioned above, it must be equal to L since vertical angles (angles formed by intersecting straight lines) are equal. Thus A equals L and we have shown that our latitude (L) is equal to the angular altitude of the NCP (A). This should work for any place on earth. In the Southern Hemisphere, the NCP becomes the SCP or South Celestial Pole.

Continuing to look at figure 11, the altitude of the intersection of the celestial equator and the observer's meridian is 90 – A, or 90 – L, or the complement of your latitude. From Washington, D.C. this would be 90 – 38, or 52 degrees. This never changes through the year or during the course of the night.

We can now examine the relation L = A. If we were to observe that the altitude of the NCP is 20 degrees, what would our latitude be? Answer: 20 degrees. Since the latitude of the North Pole of the earth is 90 degrees, what would be the altitude of the NCP as seen from the North Pole? Answer: 90 degrees. The sky would therefore look like figure 12, where the NCP is at the zenith and all diurnal paths are parallel to the horizon.

From the North Pole, nothing would rise or set during one day or night since all motion is parallel to the horizon. Over a period of one month, the moon would rise and set once. The sun would rise once per year, and stay above the horizon for six months. Half the sky would be

CROSS SECTION THROUGH EARTH AND CELESTIAL EQUATOR

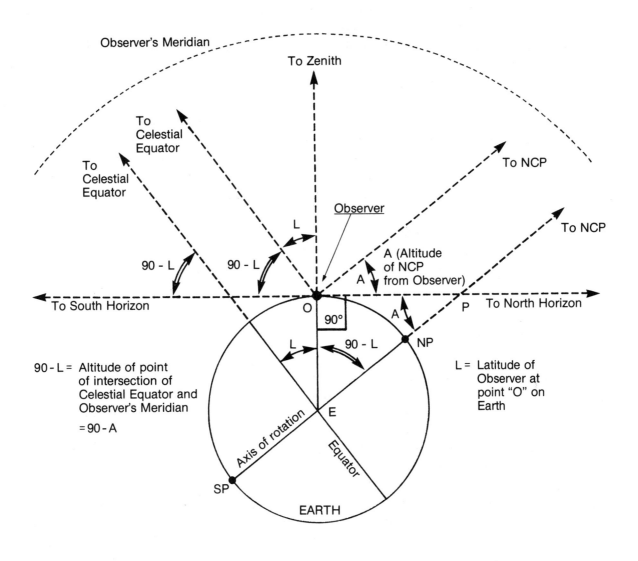

In triangle EOP:

$$90° + A + (90 - L)° = 180°$$

or $A - L + 180 = 180$

or $A - L = O$

so $A = L$

Figure 11 The fundamental relationship between the latitude of an observer and the appearance of the night sky: the altitude of the NCP equals the observer's latitude.

19

circumpolar. What part of the sky would never be seen? Where would the celestial equator be seen?

If we were to travel to the earth's equator, our latitude would become 0 degrees. Therefore, the NCP altitude would become 0 degrees too and would be literally on the north point of the horizon. At the equator, the diurnal paths are vertical (see figure 13), still parallel to the celestial equator but now perpendicular to the horizon. Thus, over the course of one day, all stars would rise and set, and none would be circumpolar.

Where is the celestial equator? What is its altitude as it crosses the observer's meridian? If the earth rotates once in 24 hours, how long will any one star be above the horizon? Questions such as these help us become familiar with the changing nature of the sky as seen from different latitudes. What can we say about the general condition for a star to be circumpolar? At the poles, all stars visible were circumpolar, but at the equator, none were, so this must be a function of latitude. Indeed, all stars within an angular distance equal to the latitude (L) of the NCP will be circumpolar. This means that all stars within 38 degrees of the NCP as seen from Washington, D.C., will be circumpolar.

Question: if a star is 15 degrees away from the NCP, what is the most southerly latitude from which it will be seen as circumpolar? Answer: 15 degrees north latitude.

Astronomical Coordinates: Declination

Astronomers locate stars by a system quite similar to the equatorial system used by navigators on the earth. They did this in order to create a system through which one might use the observed positions of stars simply and efficiently to locate the latitude and longitude of an observer on the earth. Both the celestial and terrestrial coordinate systems are based upon the earth's rotation, and therefore upon the positions of its North and South poles and the position of its equator. The purpose of this section is to introduce the celestial coordinate system and to show how it is the basis of earthly navigation by the stars.

Directly analogous to terrestrial latitude is the coordinate **declination** (DEC). Declination is measured perpendicular from the celestial equator to the NCP or to the SCP along a meridian line or, as it is called on the celestial sphere, an **hour circle.** From figure 14, a star that goes through the zenith on the observer's meridian has a declination equal to DEC. From Washington, D.C., L equals 38 degrees. Since

$$90 - DEC + L = 90,$$

the declination of the star must be equal to the latitude of the place, or 38 degrees. Thus from any latitude, we can observe what the declination of a star will be if it is seen to go through the zenith.

What about any star that crosses the meridian from our (or any) latitude and at a meridian altitude different from the zenith? First, the primary observation is the **altitude** of the object above the south point. This will be equal to the altitude of the intersection of the celestial equator and the observer's meridian ("A.C.E."), plus the declination of the star (or planet or sun or moon). Note: If the star (*) has a southern declination, it must be added

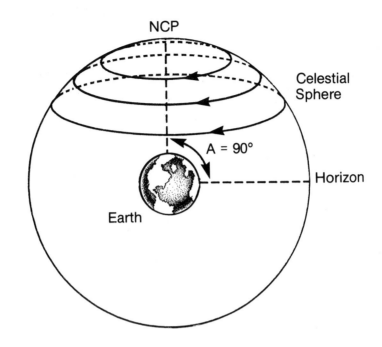

Figure 12 Diurnal paths as seen from the earth's North Pole.

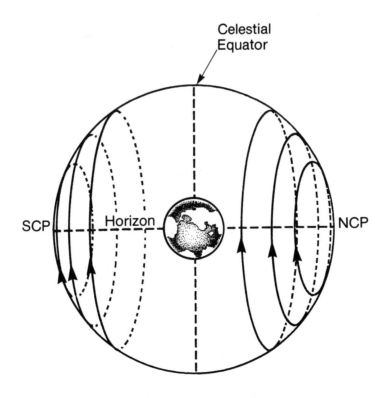

Figure 13 Diurnal paths as seen from the earth's equator.

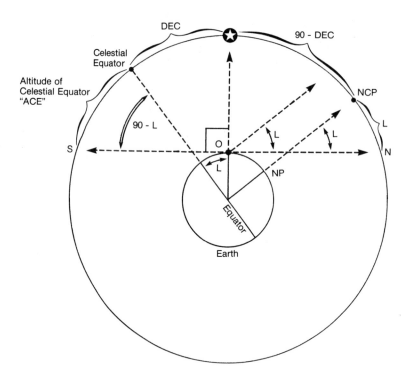

Figure 14 A star passing through the observer's zenith has a declination equal to the observer's latitude.

negatively, or subtracted. In other words:

$$\text{Alt} * = (90 - L) + (\text{DEC} *)$$

Example: at L = 38 degrees and DEC = 38 degrees, the altitude of the star (Alt *) = (90 − 38) + 38 = 90 degrees, or, the star goes through the zenith.

A second example: at L = 38 degrees and a star with declination DEC = −10 degrees, the altitude of the star will be: = 90 − 38 + (−10) = 52 − 10 = 42 degrees. Note again that if the star has a south declination, its value must be subtracted on the right hand side of the equation, rather than added.

To further examine this celestial/terrestrial relationship, we illustrate the celestial sphere again as seen in the plane of the observer's meridian (figure 15) and look at the meridian passages of three stars. The altitude of star 1 will be:

$$\text{ALT} (*1) = (90 - L) + (-\text{DEC}) = 90 - L - \text{DEC1}$$

The altitude for star 2 will be:

$$\text{ALT} (*2) = (90 - L) + (\text{DEC}) = 90 - L + \text{DEC2}$$

And for star 3:

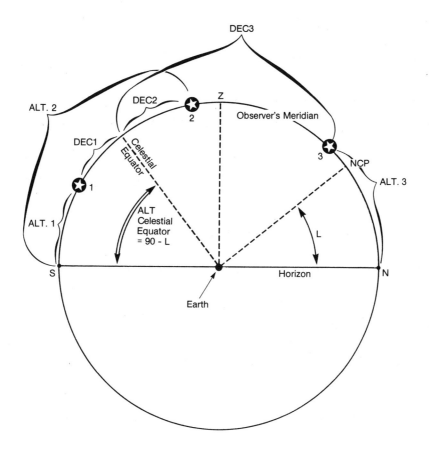

Figure 15 Three stars with different declinations have three different meridian passage altitudes.

$$Alt\ (*3) = 90 - L + DEC3.$$

From figure 15, it is clear that the altitude of star 3 will be greater than 90 degrees. So here, subtract the entire term from 180 degrees:

$$180 - (90 - L + DEC3) = 90 + L - DEC3$$

This will work equally well for calculating the meridian-passage altitudes of the sun, planets, and moon.

Right Ascension

The other celestial coordinate along with declination required to locate objects on the celestial sphere is called **right ascension,** or R.A. Measured parallel to terrestrial longitude and sometimes called celestial longitude, it is based on the position of the **vernal equinox,** a special point on the celestial equator to be discussed later. The R.A. of an object is its angular distance (in units of time or angular measure) eastward from the vernal equinox, as measured along the celestial equator.

Questions

1. If the maximum northerly declination of the sun is 23½ degrees, will it ever be seen at the zenith from Washington, D.C.? Where will it be possible to see it at the zenith (just the latitude)?

2. At the North Pole, what it the altitude of a star whose declination is 45 degrees? 20 degrees? −20 degrees?

3. What is the largest declination that is circumpolar from Washington, D.C.?

4. What is the smallest declination that is circumpolar from Washington, D.C.?

5. Derive a general relationship that can predict the range of declinations that will be circumpolar from any latitude L.

6. What is your latitude if a star, whose declination is known to be 45 degrees, passes through your zenith?

7. If you were a navigator at sea, what would be the most direct way to estimate your latitude?

2   Time

The measurement of time is an activity we all take for granted. Merely look on the wall for a clock or check your wrist watch. The type of time one determines in this manner is best known as "civil" time. It is a method of reckoning time for "civil" or social purposes, and is only an approximation that averages out seasonal variations that cause differences in the astronomical reckoning of time—the classical method of checking our watches.

Time is measured with respect to the rotation of the earth. To measure the earth's rotation, one must locate celestial objects that don't partake in its rotational motion. The sun and stars can be used, though their use will provide us with different rates of rotation for the earth. The sun will provide us with **solar time** and the stars with **sidereal time.**

Solar Time

The position of the sun in our sky is a reasonably good indicator of time. If the sun is rising, it is morning. This might be near 6 a.m. if the season is spring or fall. But in the winter, the rising sun from the mid-northern latitudes on the earth is seen around 7:30 a.m. and in the summer, close to 4:30 a.m.

At any part of the year, when the sun is due south in our sky, the time is close to noon. In the spring and fall, when the sun sets, it is 6:00 p.m. Appropriate corrections are required for winter and summer that will be covered soon. Note here that we are being careful to designate the type of hour—i.e., a.m. and p.m. What do the abbreviations mean? They stem from Latin phrases that read:

A.M.—Ante Meridian, or "before the Meridian"

P.M.—Post Meridian, or "after the Meridian"

Time then, in this manner of reckoning, must have something to do with the position of the sun in the sky with respect to the observer's meridian.

Figure 16 illustrates the apparent paths of the sun in the various seasons as an observer would see them while facing due south and looking directly at the observer's meridian. The winter sun passes lower across the observer's meridian than does the spring or summer sun. Further, the winter sun rises south of east and sets south of west, while the spring (or fall) sun rises and sets at the cardinal points. The summer sun rises and sets north of the cardinal points.

Since the angular rate of rotation of the earth is constant, the daily angular motion of the sun resulting from the earth's rotation must be roughly constant too. Because the summer path of the sun is longer than the winter path, the sun must be visible for a greater amount of time in the summer than in the winter. This result comes from the simple relationship: distance = velocity × time. For equal velocities (either angular or linear), a longer time is required to traverse a longer distance.

Local Apparent Time

We have been using the sky, specifically the position of the sun with respect to our horizon and the observer's meridian, as a clock to reckon time. Since childhood, we have all been familiar with the term "high noon"—either from direct experience or from the title of the famous western movie.

What does the term imply? Simply, at "noon" the sun appears to be highest in the sky. This can be seen easily from figure 16, where, halfway between rising and setting (morning and evening), the sun is crossing the observer's meridian, is occupying the highest portion of its daily path, and is at "high noon." This gives us the all-important idea that the passage of time and the passage of the sun are one and the same thing.

If we examine this a bit closer, we will see that the actual position of the sun with respect to the observer's meridian will allow for the determination of **local apparent time**—or time as measured by the apparent position of the sun. Local apparent time (or L.A.T.) is defined as follows:

L.A.T. = hour angle of apparent sun + 12 hours

To understand this definition, we must understand what an **hour angle** is. Most students feel that it is "our angle"—which is not far off. It is the angle the sun (or any celestial object) apparently makes with "our" observer's meridian. This angle is measured along the celestial equator, from the sun's position to the intersection of the celestial equator and the observer's meridian. This measurement is always in an eastward direction from the sun and can be anything from 0 to 360 degrees, or in time units, from 0 to 24 hours.

As an example, if the sun is on the meridian, its hour angle with respect to the observer's meridian is 0 hours, and L.A.T. is 12 hours. If the sun is one hour before the meridian (or in the southeastern part of the sky), its hour angle will be 23 hours (since you have to measure east, or in this case, away from the meridian initially) and the L.A.T. will be:

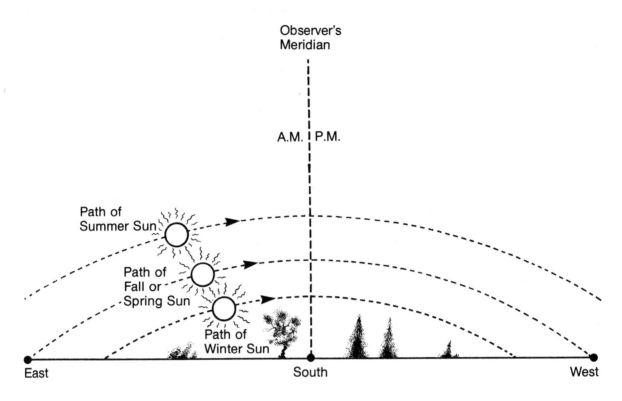

Figure 16 Paths of the sun during different seasons. The visible path length of the sun, and hence the time it spends above the observer's horizon, is longest in the summer and shortest in the winter.

$$\text{L.A.T.} = 23 \text{ hours} + 12 \text{ hours}$$
$$= 35 \text{ hours or } (35 - 24) = 11 \text{ hours}$$

since time reckoning on our system is measured in 24 hourly intervals. The period of 35 hours is therefore 11 beyond 24 or 11 hours into a new cycle.

For a third example, calculate the L.A.T. when the sun is one hour west of the meridian. Here its hour angle is one hour and so:

$$\text{L.A.T.} = 1 \text{ hour} + 12 \text{ hours}$$
$$= 13 \text{ hours}$$

This is true on a 24-hour clock. It would be 1 p.m. on a 12-hour clock.

In figure 17, the lines perpendicular to the celestial equator are called hour circles, and are circles like meridians of longitude on earth. Hour circles pass from the South Celestial Pole, through the celestial equator, and through the North Celestial Pole. Just as all places on earth on any one meridian line have the same longitude, any celestial object on the same meridian or hour circle as another one will have the same hour angle. If we look "down" on the celestial sphere from the direction of the NCP (figure 18), we see a projection of the celestial equator as the periphery of the celestial sphere with the NCP at the center. Hour angle is measured eastward from the object to the observer's meridian.

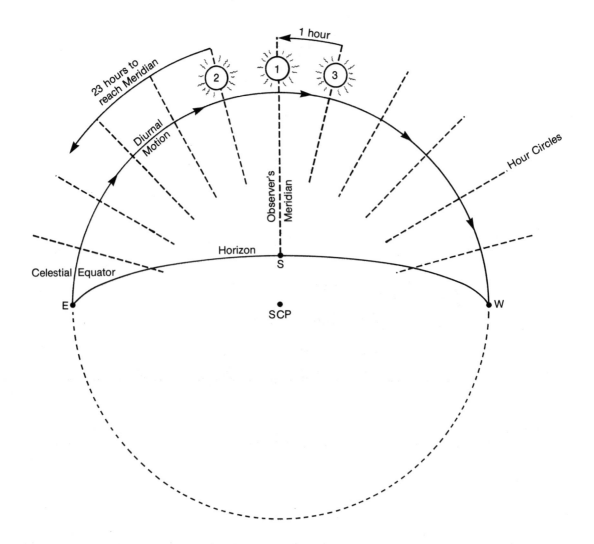

Figure 17 Looking south: a distorted view of the celestial equator, extending 360 degrees around the sky showing "hour circles" and the sun in three positions—at 11 hours L.A.T., 12 hours L.A.T., and 13 hours L.A.T. (positions 2, 1 and 3 respectively).

The hour angle of any object on the celestial sphere, but not necessarily on the celestial equator, is determined by noting which hour circle passes through the object and finding the hour angle of the intersection of that hour circle when it crosses the celestial equator, as figure 19 illustrates.

In figure 19, object #1 has an hour angle of one hour, but so does object #2, because it is on the same hour circle. Its celestial longitude (or as astronomers call it, its right ascension) is the same as that for the object on the celestial equator. Any two objects on the same hour circle have the same hour angle at any one time, and the same right ascension at all times.

Now that we have examined the concept of hour angle, we must see how L.A.T. is related

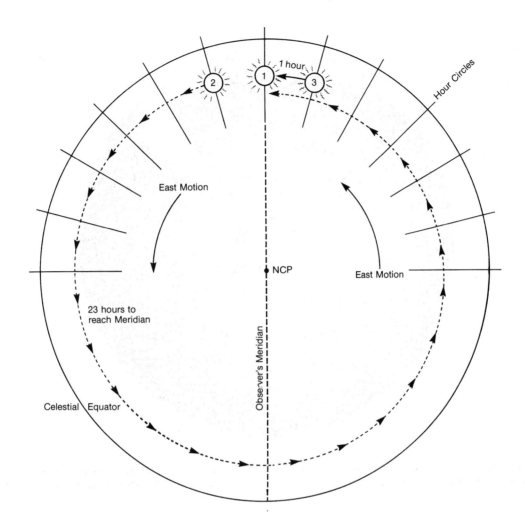

Figure 18 The celestial sphere as seen from an imaginary point above the NCP. The outside circle is the celestial equator, the vertical dashed line through the NCP is the observer's meridian. The hour angle of sun #2 is 23 hours; sun #1 is 0 hours; sun #3 is 1 hour. All counterclockwise motion is "eastward" motion, all clockwise motion is "westward" motion.

to civil time or "Standard Time"—the time read by the average clock on the wall.

As we will see in the following sections, because of the earth's orbital motion around the sun, an earthly observer will see the sun apparently travel among the stars. This motion causes solar time to differ from sidereal time each day by a small but cumulative amount. Since the sun travels on a path inclined to the celestial equator, and since the earth's motion around the sun is not uniform, the daily motion of the sun among the stars is not uniform either. The correction for these non-uniform effects is called the **equation of time,** which will not be covered in these lectures. Another correction needed to convert L.A.T. to standard time is a correction for longitude on earth, which will be discussed.

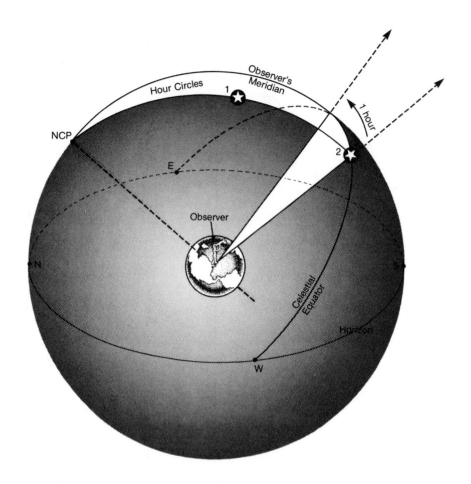

Figure 19 The hour angles of two objects on the same hour circle are always equal. In this illustration, the hour angles of objects 1 and 2 are both one hour, even though one is on the celestial equator and the other is much farther north in the sky.

Standard Time

The earth is divided into standard zones of time reckoning for many reasons. Since time is somehow measured with respect to the observed position of the sun as seen from one particular point on earth, different parts of the earth experience different times because the sun will be seen in different parts of the sky from places separated in longitude. When the sun is seen rising from New York, Los Angeles is still in darkness and it is lunch time in England.

As the earth rotates eastward, different points on the earth's surface approach the sun (those where the sun is seen in the eastern part of the sky) and others recede from the sun (where the sun is seen in the western sky). In figure 20, the sun is seen just rising at point #1. At point #2, the sun is well above its horizon and approaching its observer's meridian from the east. At point #3, the sun is on the meridian, and at point #4, the sun is well past the meridian and about to set in the west (point #5).

Since society somehow decided that there are to be 24 hours of time during one daily cycle of the earth, it is logical that the earth be divided into 24 zones of time reckoning so each zone covers one hour of time. Since there are 360 degrees of longitude, each zone contains approximately 15 degrees of longitude (360/24 = 15). Variation within these zones is needed to allow regions bounded by political borders (like states, counties, and cities) to have the same time.

Within each zone, all people measure time by the same standard clock on the "standard meridian" that defines that zone. These standard meridians, spaced every 15 degrees around

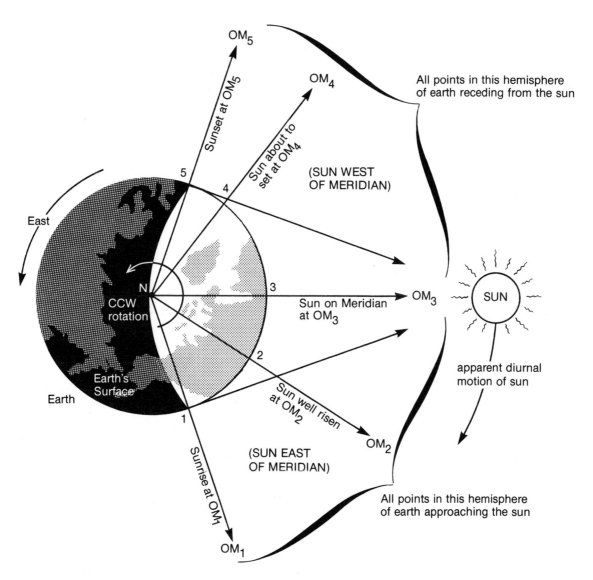

Figure 20 The varying position of the sun. Five points on the earth's surface separated in longitude see the sun in five positions from sunrise to sunset.

the earth, begin through universal agreement at Greenwich, England. Most points on the eastern seaboard of the United States are close to the 75 degree W longitude meridian and therefore read standard time as it is "seen" from that meridian which is 75 degrees/15 degrees = 5 time zones away from Greenwich. Washington, D.C., is in this zone (the eastern standard time zone) and therefore experiences time five hours behind Greenwich. People in Los Angeles are close to the 120-degree meridian and therefore are 120 degrees/15 degrees = 8 time zones behind Greenwich and three zones (or hours) behind the eastern standard time zone.

Of course, not everyone within any one zone is on the same meridian of longitude. Therefore, each person's L.A.T. is not necessarily equal to standard time. Only those people living on the 75th meridian will have their L.A.T. = standard time (neglecting equation of time variations). For everyone else, a correction for longitude is needed and is a function of the difference between the observer's longitude and the standard meridian longitude. The longitude of Washington, D.C., is approximately 77 degrees west. We are, therefore, 2 degrees west of our standard time meridian of 75 degrees. Since the earth takes 24 hours to rotate 360 degrees, in one hour it will rotate 360/24 or 15 degrees, or one degree every four minutes (60 minutes/15). For a difference of two degrees west, then, we will be eight minutes behind our standard meridian, thus our L.A.T. is different from standard time by eight minutes (again neglecting the equation of time).

If we were at 72 degrees longitude west, we would be three degrees ahead of our standard meridian and therefore would see things happening (like meridian transits of the sun and stars) 12 minutes before people would who are sitting on the standard meridian. Our L.A.T. would be 12 minutes later than standard time.

To review: observers at these three longitudes experience 12:00 EST at the same time but experience 12:00 L.A.T. at different times. At 12:00 L.A.T. as seen from 75 degrees, L.A.T. will be 12:12 at 72 degrees west and 11:52 at 77 degrees west.

By turning this discussion around, we can see how knowing L.A.T. and having a clock that faithfully reads standard time can help us determine longitude (neglecting the equation of time). If we are on a ship at sea and note that our local apparent sun is transiting the meridian (L.A.T. = 12:00), but our standard time clock reads 11:20, we should immediately realize that our L.A.T. is 40 minutes ahead of the clock and so we are 40/4 = 10 degrees east of 75 degrees, or are sitting at 65 degrees west longitude. Recall that the earth rotates four minutes per degree of longitude. This is where the "4" comes from in the previous example.

Questions

1. What is your longitude if your local sun is seen transiting your local meridian and a standard clock reads:
   A. 12:00 EST B. 12:00 EDST C. 11:40 EST D. 12:16 EST

2. What is your longitude if your L.A.T is 14 hours and your standard clock reads:
   A. 12:00 EST B. 12:00 PST C. 11:00 EST

Sidereal Time

Thus far we have been considering time as determined by the position of the sun. Since the sun apparently moves among the stars each day, time reckoned by the stars will differ from time as determined from the sun because their respective rates of passage across our meridian will be a bit different.

We must recall that the earth revolves around the sun in approximately 365 days. Therefore, in one day it moves roughly 360 degrees/365 days, or approximately one degree, around the sun. The sun moves eastward among the stars, so this one degree of daily motion means that it "lags" behind the stars by the amount of time it takes the earth to rotate one degree, or four minutes. To see this, we depict the orbit of the earth as seen from a position in space approximately above the North Pole of the earth but not exactly (see figure 21). From the earth at position #1, the sun is on the meridian and is superimposed on an arbitrary point on the celestial sphere. At position #2, after one complete rotation of the earth with respect to the stars on the celestial sphere, that same arbitrary point on the celestial sphere is again on the meridian, but the sun is still in the eastern part of the sky (as determined by the counterclockwise rotation of the earth, which is defined as eastward).

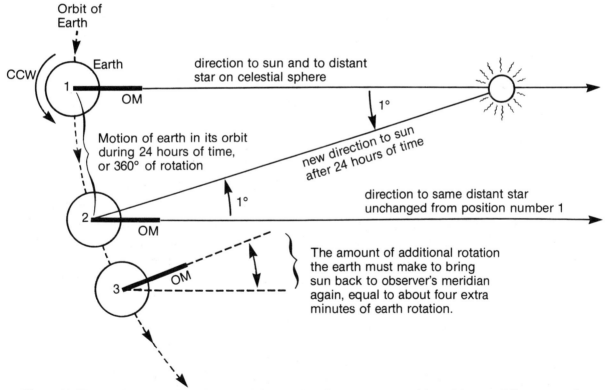

Figure 21 The earth in its orbit during one daily rotation. The apparent position of the sun shifts eastward with respect to stars by about 1 degree per day. Successive transits of the sun across the observer's meridian lag by 1 degree or 4 minutes per day behind successive transits of stars across the observer's meridian. After 360 degrees of rotation, the earth is at position 2, and moves to position 3 in the four additional minutes needed to bring the sun across the observer's meridian again.

Since the earth has moved about one degree in its orbit around the sun, its line of sight to the sun at position #2 is a **transversal** to the set of parallel lines to the arbitrary point on the celestial sphere, and therefore the sun is one degree east of the meridian. The earth must therefore rotate an extra degree to bring the sun back to the meridian (position #3), which takes an extra four minutes. If the sidereal day is defined as 24 hours, the solar day will therefore be 24 hours 4 minutes. We commonly designate the solar day as 24 hours of solar time. Therefore, the sidereal day becomes 23 hours 56 minutes of solar time.

Sidereal time is measured much the same way as solar time, but instead of using the sun, a specific point on the celestial sphere called the **vernal equinox** is used. This point is where the sun passes across the celestial equator, moving from the southern to the northern heavens, which occurs around March 21 and is the designation of the first day of spring. The vernal equinox is the base point of our celestial coordinate system (like Greenwich on earth) and has a declination (or celestial latitude) of 0 degrees and a right ascension (or celestial longitude) of 0 hours. Sidereal time is defined in the following way:

Sidereal Time = Hour Angle of the Vernal Equinox

Hour angle is, of course, measured in the same way here as it was for the sun or for any object on the celestial sphere. If we wish to determine the sidereal time, we must know where the vernal equinox is. If the date is around March 21, the sun is on or near the vernal equinox (see figure 22), and therefore the sidereal time will be similar to L.A.T. on that date with a twelve-hour correction:

On March 21: Hour Angle of Sun = L.A.T. − 12 Hours
= Hour Angle of
Vernal Equinox,

or the hour angle of the sun equals the sidereal time, which also equals the L.A.T. − 12 Hours.

On September 21, six months from the vernal equinox, the sun has moved to the autumnal equinox, or that part of its path where it is crossing the celestial equator from the north to the south. This six month difference means that the sun's right ascension must be 180 degrees, or 12 hours different from 0 degrees. Thus the sun's right ascension is 12 hours on the autumnal equinox. On this date, L.A.T. is equal to sidereal time. If the sun is found on the meridian, 0 hours right ascension (or the vernal equinox point) will have an hour angle of 12 hours, so the sidereal time must be 12 hours. Thus on September 21:

Hour Angle of Sun = Sidereal Time − 12 hours
or
Sidereal Time = H.A. Sun + 12 hours
Thus at noon (L.A.T. = 12 hours), the sidereal time will be 12 hours on
September 21, and we may write:
L.A.T. = Sidereal Time

Be careful to note that these simple relationships hold only for the two equinox dates. For any other dates, simple addition can be used to find the approximate conversion: since the sun

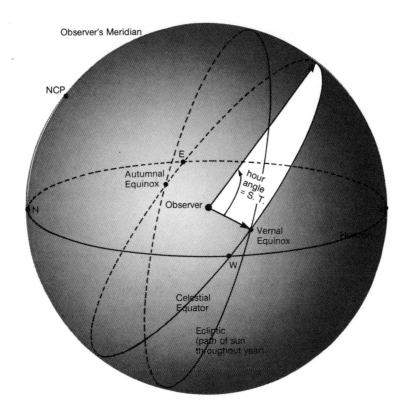

Figure 22 The celestial sphere showing the vernal equinox and the autumnal equinox and the measurement of sidereal time, or the hour angle of the vernal equinox. The path of the sun during the year with respect to the stars is called the ecliptic.

moves 24 hours of right ascension in 12 months, it must move, on the average, two hours of right ascension in one month on the ecliptic. Thus, one month after the vernal equinox (April 21), the sun must have moved from right ascension 0 hours to right ascension 2 hours and be two hours behind (or east of) its position on the vernal equinox. Therefore, if sidereal time was equal to the hour angle of the sun on the vernal equinox, one month later it must be two hours behind sidereal time, or the hour angle of the sun equals the sidereal time minus 2 hours on April 21, or:

$$\text{H.A. Sun} = \text{Sidereal Time} - 2 \text{ hours}$$

and since at L.A.T. noon the hour angle of the sun equals 0 degrees, then the sidereal time will equal 2 hours.

The use of sidereal time is of great importance in finding astronomical objects at night and also for navigation by the stars. The right ascensions of stars do not change noticeably within years, decades, or even centuries of time, and so to determine where they might be seen at some time of night and date of the year, one only needs to convert their celestial positions to

35

their terrestrially oriented hour angle position, or the angle the object makes with respect to the observer's local meridian. One simple equation suffices to do this using the right ascension of the object, its hour angle, and the sidereal time:

Hour Angle (of star, = Sidereal Time − Right Ascension (of object)
planet, sun)

or,

$$\text{H.A.} = \text{S.T.} - \text{R.A.}$$

As an example, find the sun in the sky on March 21 at L.A.T. = 12 noon: R.A. sun = 0 hours on that date. At L.A.T. = 12 noon, where is it in the sky?

$$\text{H.A. (Sun)} = 0 \text{ hrs.} - 0 \text{ hrs.} = 0 \text{ hrs.,}$$

or the sun is on the meridian. But we knew that already. Let's try another more realistic example: If the star "NASM" has a right ascension of 6 hours, what is its hour angle on March 21 at L.A.T. = 20 hours? Solution: on March 21 R.A. sun = 0 hours, so at L.A.T. = 20 hours, the H.A. of the sun is 20 − 12 = 8 hours. In the equation

$$\text{H.A. Sun} = \text{S.T.} - \text{R.A. (Sun)}$$

substitute:

$$8 \text{ hr.} = \text{S.T.} - 0 \text{ hrs.}$$

The sun's R.A. is = 0 on this date, and S.T. = 8 hours so now for our star "NASM":

$$\text{H.A. "NASM"} = 8 \text{ hr.} - \text{R.A. "NASM"}$$

or

$$= 8 \text{ hr.} - 6 \text{ hr.}$$
$$= 2 \text{ hr.,}$$

or two hours west of the observer's meridian.

If the declination of the star "NASM" is 0 degrees (on the celestial equator), it will be seen in the southwest about 45 degrees above the horizon. This shows roughly how astronomers are able to find objects in the sky using only their coordinates.

When will the star "NASM" cross our meridian on March 21? This is useful information because objects are best seen when they are highest in the sky or crossing the meridian. First, the hour angle of "NASM" must be 0 hours to cross the meridian, so,

$$0 \text{ hours} = \text{S.T.} - 6 \text{ hours}$$

or at 6 hours sidereal time, the star will be crossing our meridian. To convert to L.A.T. on this date, recall that the R.A. of the sun is 0 hours on March 21, so

$$\text{H.A. Sun} = 6 \text{ hours} - \text{R.A. Sun}$$

or

$$\text{H.A. Sun} = 6 \text{ hours} - 0 \text{ hours}$$
$$= 6 \text{ hours}$$

But the hour angle of the sun is L.A.T. − 12 hours, so L.A.T. = 18 hours or 6 p.m. Therefore, the star "NASM" will be crossing the meridian at 6 p.m., or close to sunset.

The constellation Orion has an approximate right ascension of 6 hours. When will be the best time (i.e. on the meridian due south) to see it early in the evening just after sunset? Answer: In the spring. This can be checked by looking at the above example for the star "NASM" or consulting a movable star-finder chart that embodies many of the calculations we have been doing here.

Questions

1. At approximately what season will the constellation Orion be seen rising in the east (H.A. = 18 hours) at sunset?

2. During what season will the constellation Orion not be visible at all at night?

3. If the planet Venus is two hours east of the sun, will it be seen in the morning or evening sky? If the date is March 21, at what time will Venus transit the meridian?

4. On November 21, what is the R.A. of the sun. What will the sidereal time be when the sun is on the meridian?

3 The Revolution of the Earth Around the Sun

We now take a look at the second motion of the earth: its orbital motion around the sun, or its **revolution.** As discussed in the last chapter on time, the sun apparently moves eastward among the stars by an amount averaging one degree per day. This is due to the earth's revolution around the sun, which causes the earth, sun, and stars to assume new positions with respect to each other constantly. We mentioned earlier that the sun's eastward motion (or lag) was not constant from day to day. In some parts of the year, the lag is greater than one degree per day, and in others, it is less. This is due to two conditions:

1. The orientation of the earth's axis with respect to the direction of its motion around the sun.

2. The acceleration of the earth's motion around the sun due to the ellipticity of the earth's orbit, as described by Kepler's second law of planetary motion (see chapter 8).

    The first condition is the primary cause of the seasons, the second is a minor seasonal variation. We will discuss the first condition here, and leave the second for a later time. The combination of these two periodic variations is called the Equation of Time, as noted before. Accurate navigation techniques require that it be accounted for, but this is beyond the purpose of this book. For a full discussion, see William Smart's *Text-Book on Spherical Astronomy* or George Abell's *Exploration of the Universe,* listed in the Bibliography.

The Earth's Orientation in Space

In the planetarium, or after several weeks of observing the altitude of the sun at noon with an **astrolabe,** it should be clear to us that the solar altitude at noon does not remain constant (see

Appendix). In the spring months, the noon solar altitude will increase and in the fall, the altitude will decrease. After one complete year of observations have been collected, we find that the changing altitude of the sun will trace out a wave if plotted on a piece of graph paper whose coordinates are meridian altitude and time.

Plotting the solar altitude at noon through the course of a year, as seen from Washington, D.C., or literally from any place in the mid-northern latitudes, will produce a chart similar to that in figure 23.

In December, the sun's altitude on the observer's meridian is less than 30 degrees, while in March or September, its altitude is about 52 degrees, and in June, its altitude is a maximum at about 75 degrees, all as seen from Washington, D.C., of course.

What could cause such a seasonal variation in the sun's noontime altitude? Somehow, the sun must change in position with respect to our celestial coordinate system, specifically, to the celestial equator. As we have already seen in chapter 1, the meridian altitude of the celestial equator is always 90 degrees – latitude, and hence does not change seasonally. The sun does change its altitude on the celestial meridian during the year, and therefore its coordinate position, its declination, must change. Over one year of time, the sun's declination changes by 47 degrees, moving from a low of 23½ degrees below the celestial equator to a high of 23½ degrees above the celestial equator (see figure 23, the vertical coordinate on the right-hand side).

If we were to create a model of the earth-sun system that would describe the sun's observed seasonal variation, it might look like figure 24 based upon the earth as center. This view was indeed believed for thousands of years to be the correct model. During the 16th and 17th centuries, however, largely through the efforts of Copernicus, our modern heliocentric, or sun-centered, view came into being. Thus the picture was changed to that in figure 25. Here

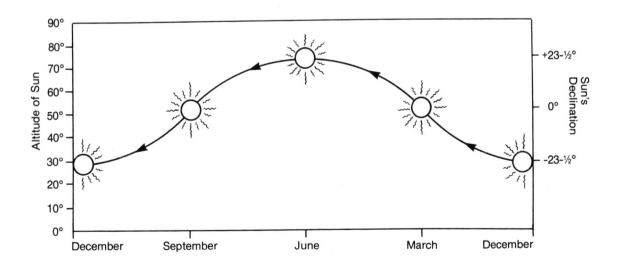

Figure 23 Changing altitude of the sun as seen at L.A.T. = 12 hours throughout a year.

we see the earth traveling around the sun, but always keeping the direction of its axis of rotation aligned in the same direction with respect to the stars. The earth's axis of rotation is tilted with respect to its direction of motion around the sun.

In the summer, the sun is seen above the celestial equator, and the north pole of the earth is totally in sunlight. In one day's time, the earth's motion around the sun is barely one degree, so during one complete rotation of the earth on its axis during the summer, the north end of the axis will constantly be in sunlight and the south part of the axis will constantly be pointed away from the sun and will therefore be in darkness. All points in the Northern Hemisphere will see the sun above the celestial equator.

In the winter, just the opposite occurs. The sun will be below the celestial equator as seen by northern observers, and the North Pole will constantly be in darkness. In the fall and spring, the sun will be on or near the celestial equator, and all parts of the earth will experience nearly equal amounts of sunlight and darkness each day.

To see how this orientation affects the daily motion of the sun across our sky (produced by the rotation of the earth) we must return to the old geocentric concept and construct a celestial sphere with the sun's path, the celestial equator, and the observer's meridian all centered on the earth (see figure 26). Here we see the sun above the celestial equator. Its daily (or diurnal) path westward is marked, as is its declination (DEC), its hour angle (H.A.), and its right ascension (R.A.).

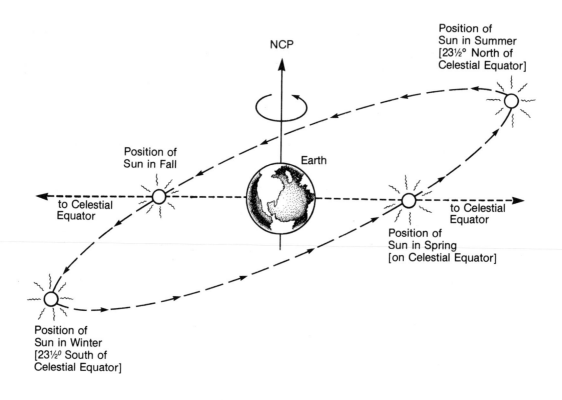

Figure 24 Geocentric view of path of sun with respect to the earth's celestial equator.

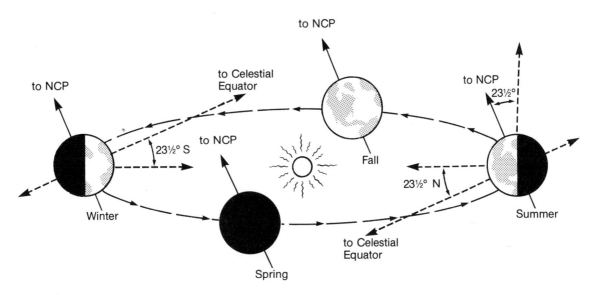

Figure 25 Heliocentric view of the orbit of the earth, showing changing position of the sun with respect to the earth's celestial equator.

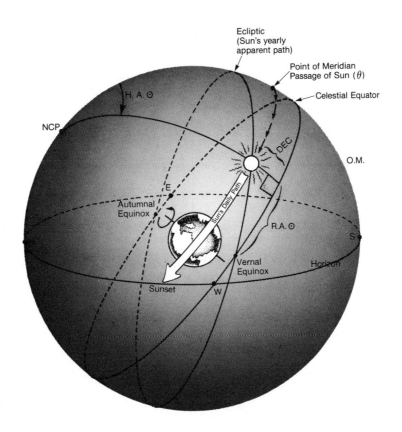

Figure 26 The position of the sun in the afternoon as seen on a day in late spring. Note its northern declination (DEC), its hour angle (H.A.), and its right ascension (R.A.). On this day the sun transits the observer's meridian north of the celestial equator and sets north of the west point of the horizon.

The sun's daily path is a small circle above the celestial equator that rises north of east and sets north of west. The altitude of the meridian passage of the sun's diurnal circle is greater than the meridian passage of the celestial equator. With the sun above the celestial equator, the season is late spring. The point marked vernal equinox is the conjunction of the celestial equator and the sun's path (the ecliptic) as the sun moves northward in its yearly path. On this date, what would the sun's path look like as you stand looking south (see figure 27)? We see that the meridian passage altitude of the sun is greater than the meridian passage altitude of the celestial equator. Question: What is the altitude of the sun? It appears to be:

Altitude of sun's
passage across meridian = altitude of meridian passage of celestial equator
+
declination of the sun on that date

Let the declination be 10 degrees north on this date (for any date in the spring or summer, it must be between 0 and 23½ degrees north), then the altitude of the sun when it crosses the meridian must be $90 - L + 10$ degrees. From Washington, D.C., L is 38 degrees, so $90 - L = 52$ degrees, and the altitude of the sun is, therefore, $52 + 10$ degrees $= 62$ degrees.

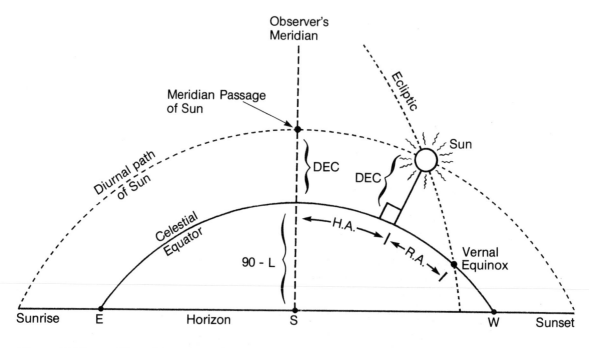

Figure 27 The position of the sun on the same day at the same time as in figure 26, but now seen by an observer facing south in the Northern Hemisphere.

Let us now examine the altitude of the sun as it crosses the meridian on four dates: the spring and autumnal equinoxes and the summer and winter solstices. The following table summarizes all pertinent data:

Date	Declination of Sun	Latitude
March 21 Spring (Vernal Equinox)	0	38
June 22 Summer Solstice	23 ½	38
September 21 Fall (Autumn Equinox)	0	38
December 21 Winter Solstice	− 23 ½	38

We use the same relationship as before:

Alt. of sun on meridian = 90 − L + declination of sun

and find:

March 21: Altitude = 90 − 38 + 0 = 52 degrees
June 22: Altitude = 90 − 38 + 23 ½ = 75 ½ degrees
Sept. 21: Altitude = 90 − 38 + 0 = 52 degrees
Dec. 21: Altitude = 90 − 38 + (− 23 ½) = 28 ½ degrees

The declinations of the sun can be found from the graph (figure 23) earlier in this section. For any date, we can use this graph to find the meridian altitude of the sun.

Example: In mid-August, what is the altitude of the sun as it crosses the meridian as seen from Washington, D.C.?
Answer: 90 − 38 + (about +15 as read from graph in figure 23)
    = 67 degrees.

If the observation was made from Pensacola, Florida (L = 30 degrees), what would be the altitude?
Answer: 90 − 30 + 15 = 75 degrees.

At what latitude would the sun be seen on the zenith on this date?
Altitude of the sun at zenith = 90 = 90 − L + 15, or, solving for L, we have:
L = 15 degrees (or the north coast of South America).

Questions

Using figure 23, and the discussion above, you should be able to answer the following questions:

1. On May 15, you observe the meridian altitude of the sun to be 45 degrees, what is your latitude? If the standard time clock on the wall reads 12:04, EST, what is your longitude?

2. Can the sun cross the meridian at the zenith from Washington, D.C.? If not, why not?

3. What is the northernmost latitude in the Northern Hemisphere that can see the sun at the zenith? What is this place called?

4. On what date(s) will the sun be at the zenith as seen from the earth's equator?

5. On June 22, what will be the altitude of the sun as seen from a latitude equal to Washington, D.C., but in the Southern Hemisphere?

Land of the Midnight Sun

As we have seen in previous sections, the altitude of the North Celestial Pole (NCP) is equal to the observer's latitude. As we travel north, the NCP should rise. Since the meridian passage of the celestial equator is $90 - L$, as L gets greater, $90 - L$ will decrease. When L = 90 degrees, for example, $90 - L = 0$ degrees, and the altitude of the celestial equator will be 0, and the celestial equator will be seen on the horizon. There is a minimum northerly latitude where, at one time of the year, the sun never rises or sets during the course of a day. To find this latitude, we look back to the relationship:

$$\text{Alt. sun on meridian} = 90 - L + \text{declination of sun.}$$

The southernmost declination of the sun occurs on December 22 and is $-23\frac{1}{2}$ degrees, therefore, on that date the solar altitude is:

$$90 - L - 23\frac{1}{2} = 66\frac{1}{2} - L.$$

For the sun not to rise, its altitude must be zero or less at all times during the day. Letting it be zero:

$$0 = 66\frac{1}{2} - L$$

or

$$L = 66\frac{1}{2} \text{ degrees north}$$

Therefore, we see that on December 22, the meridian altitude of the sun is 0 degrees from latitude $66\frac{1}{2}$ degrees north, and the sun never rises on that day. This latitude is called the Arctic Circle.

The Arctic Circle is a most interesting place in the summer months. Anyone who has spent summers in far northern latitudes knows that twilight can last well past 10 p.m. and even until 11 p.m. (in Scotland, for example). On June 22, the declination of the sun is $+23\frac{1}{2}$ degrees, and from the Arctic Circle, the meridian altitude of the sun is:

$$90 - 66\frac{1}{2} + 23\frac{1}{2} = 47 \text{ degrees}$$

However, from this latitude, all celestial objects within L degrees of the NCP are circumpolar. Since L = $66\frac{1}{2}$ degrees, all objects with $90 - 66\frac{1}{2} = 23\frac{1}{2}$ degrees declination or greater are circumpolar. Therefore, the sun will be circumpolar (just barely) on this date. It will transit the southern meridian at 47 degrees and then travel around the sky (in its apparent diurnal path), barely grazing the northern horizon as it crosses the northern meridian.

If the observer's latitude is greater than $66\frac{1}{2}$ degrees, the restraints for this "midnight sun" will be further eased. From latitude 75 degrees north, all objects with declinations greater than 15 degrees (90 − 75) will be circumpolar, and hence the sun will be circumpolar when its declination is greater than 15 degrees. Looking at the graph again (figure 23), we see that the sun's declination is greater than 15 degrees between mid-May and mid-August. So during the interval between these dates, the sun never sets from latitude 75 degrees north or from points farther north.

Questions

1. On what date does the sun rise as seen from the North Pole?

2. On what date does the sun set as seen from the South Pole?

3. During which months would the sun never be seen from 75 degrees north latitude?

4. From what latitude would the sun be seen peeking over the northern horizon on June 22?

Azimuths of Rise and Set

We now ask: where on the horizon does the sun rise and set at different times of the year? In ancient times, right up to the present, azimuthal risings and settings of the sun foretold the seasons just as we have seen its meridian altitude could do. Earlier, we have seen that the sun does not necessarily rise due east and set due west at all times of the year. We saw that in the summer months, the sun's diurnal path rises north of east and sets north of west. In the same way, we might be able to represent the sun's winter path by a small circle south of the celestial equator. The small diurnal circle would rise south of east and set south of west, as seen in figure 28.

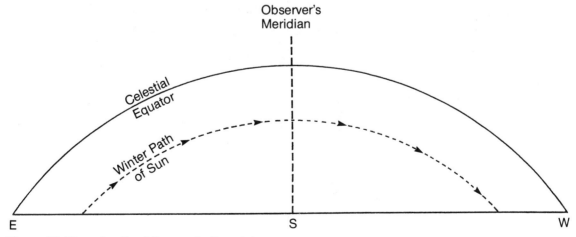

Figure 28 Diurnal path of the sun in the winter.

45

If we compare the winter path of the sun, as depicted in figure 28, with its summer path as seen in figures 26 and 27, we immediately see that the length of the diurnal path of the sun above the horizon is smaller in winter than in summer. This means that the sun is above our horizon for a shorter time in winter than in summer. Note too that the meridian altitude of the sun in winter is lower than in summer. This correlation with the seasons is not accidental. The amount of sunlight received by any part of the earth's surface *determines* the type of season we experience.

To be able to determine the actual lengths of daylight on seasonal dates during the year requires a level of mathematics not assumed here. We can, however, discuss the matter in the planetarium, or in a laboratory using transparent celestial globes. Here, we present a simplified approach.

First, we must show that from any latitude (other than at the poles), exactly half of the celestial equator is always above the horizon. Figure 29 illustrates the situation. The celestial equator is a **great circle** passing through the east and west points.

As can be seen from figure 29, the celestial equator crosses the horizon plane, which is also a great circle. Since any great circle cuts the sphere upon which it is inscribed in two equal parts, a diameter to the great circle is a diameter to the sphere. Thus, the diameter of the celestial equator great circle is a diameter for the celestial sphere. But the line joining the east and west

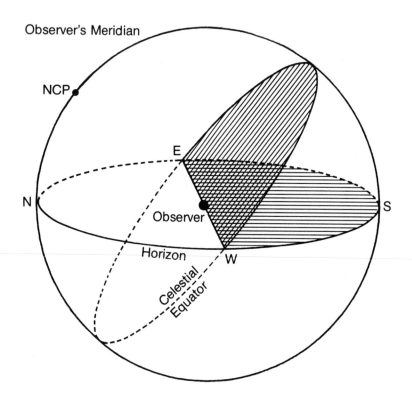

Figure 29 The intersection of the horizon plane and the celestial equator bisects both planes.

points is also a diameter to the great circle that produces the horizon. So the celestial equator and the horizon share the same diameter, the east-west (EW) line. Since any diameter to a circle cuts the circle in half, the celestial equator must cut the horizon in half and so the horizon must cut the celestial equator in half.

If we imagine the celestial sphere as a great turning clock, any point on the celestial equator will rise due east and travel exactly half its *total* diurnal path (24 hours) to the west point. Half of 24 is 12, so any point on the celestial equator will be above the horizon for 12 hours and below it for 12 hours. The sun is on the celestial equator on the equinox dates, which are so named because on those dates, everyone on earth experiences equal daylight and equal night (12 + 12 hours).

By similar arguments, we can show that any small diurnal circle inscribed on the celestial sphere parallel and north of the celestial equator will have more than half of its circumference above the horizon, and any small diurnal circle south of the celestial equator will have less than half of its circumference visible, as seen from the Northern Hemisphere. Thus, when any celestial object is on a small diurnal circle north of the celestial equator more than half its path will be above the horizon and therefore the object will be seen for a time greater than 12 hours (see figure 30), since the angular rate of motion of objects is the same, no matter the size of the diurnal circle.

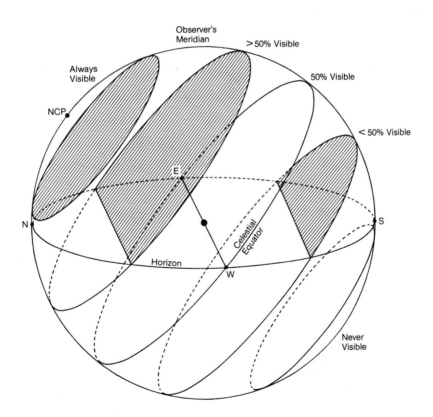

Figure 30 Northern and southern diurnal circles. Circles north of the celestial equator are more than half above the horizon. Circles below the celestial equator are less than half above the horizon.

Exercises for Planetarium or Laboratory Celestial Globe

1. Set instrument for your latitude (Washington D.C. = 38 degrees)

2. Place sun on each of four following dates:
   March 21
   June 22
   September 21
   December 22

   For each:
   a. Note the value of the hour circle crossing the meridian when the sun rises (this is actually the sidereal time of sunrise).
   b. Note the approximate azimuth of the sun at sunrise (E, NE, SE, etc.).
   c. Allow the earth to rotate until the sun is on your observer's meridian. Note its meridian altitude and the sidereal time of passage.
   d. Allow the sun to set. Note its azimuth and the hour circle passing the meridian at sunset (sidereal time).

3. Repeat for different latitudes (include Tropics, Arctic Circle, and one latitude of individual choice).

4. From your compiled data sheet, discuss how the length of daylight and the quality of sunlight vary from season to season, and for each season, how the length of daylight and its quality vary with latitude.

5. During steps 2 and 3, note which constellations are on the meridian, which are rising, and which are setting. Compile a list of constellations that should be seasonally visible. Note where each appears in the sky. On the next clear night, go out and identify constellations in the south and close to your meridian and compare this with what you have predicted.

The Seasons   (data sheet, p. 104)

We now have almost all the information we need to understand why the seasons occur as they do. Certainly the absence or presence of sunlight is an important determiner of the average temperature of any point on the surface of the earth. So, when the sun is above the horizon for a longer time, a greater amount of radiant energy will be available to heat the earth. Further, as can be seen from figure 30, when the sun's path is above the celestial equator, not only is its duration above our horizon longer than 12 hours but the altitude it attains at noon is greater than the altitude of the intersection of the celestial equator and the observer's meridian.

During the summer months, the sun is visible for more than 12 hours, and is higher in the sky than is the celestial equator and the observer's meridian. But in the winter months,

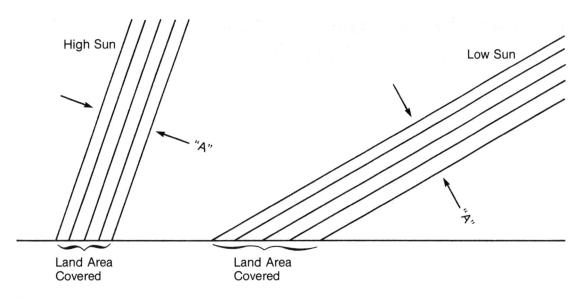

Figure 31 High sun rays fall over a smaller area than low sun rays, hence high sun rays are more concentrated on the ground than low sun rays. In both cases, the width of a ray of sunlight "A" is constant.

everything is reversed; the sun is visible for less than 12 hours and never reaches great meridian altitude.

From the previous exercise, we know that there can be as much as 15 hours of daylight on the summer solstice as seen from Washington, D.C., and on that date, the sun attains an altitude of about 77 degrees. On the winter solstice, the sun is visible for only 9 hours, and never rises higher than 30 degrees.

The length of daylight is an important determiner of the amount of energy received from the sun—but the height the sun attains is also an important factor. As figure 31 illustrates, when the sun is at a great altitude, its radiant energy is concentrated upon the surface of the earth. When the sun is at a low altitude, its rays glance the surface of the earth and are diluted. When at a low altitude, the sun's rays are spread over a wider area; hence, per unit area, not as much heating takes place.

If these two factors—height of the sun and the number of hours of visibility—were the only ones important in determining the surface temperature of the earth, then the seasonal cycle of temperature would agree exactly with the dates of the seasons. The coldest day would be December 21 and the hottest day would be June 21. But this is not true. Other very important factors must be accounted for, such as the direction and nature of the prevailing winds, the presence or absence of large bodies of water and their location with respect to the prevailing winds, the presence or absence of ocean currents, and the general nature of continental topography.

All these additional factors cause the annual temperature variations experienced by a locale to behave in a manner far more complicated than would be suggested by the simple yearly cycle of the sun. Nevertheless, the fundamental character of that cycle is driven by the seasonal changes in the position of the sun in the sky.

4 Times of Rising, Setting, and Meridian Transit for Stars, Planets, and the Moon

We have spent considerable time discussing the astronomical coordinate system from which the length of daylight can be calculated for any date and for any latitude. Very much the same techniques can be used for determining the visibility of stars, constellations, planets, and the moon, using a planetarium, celestial globe, star charts, binoculars, or telescopes.

Observations of Stars

If the right ascension (R.A.) and declination (DEC) of a star are known, then its position in the sky for any time, date, and latitude can be determined. Again, we use the basic relationship:

Hour Angle (of star, etc.) = Sidereal Time − R.A. (star, etc.)

Example: If a star's coordinates are R.A. = 6 hours, DEC = 20 degrees, on what date of the year will it be seen on the meridian at 9:00 p.m. EST?

First, if it is on the meridian, its H.A. = 0 Hours.

$$\text{Therefore H.A.} = 0 = \text{S.T.} - \text{R.A.}$$
$$= \text{S.T.} - 6 \text{ Hours}$$

and

$$\text{S.T.} = 6 \text{ hours.}$$

We already knew this answer because the value of the hour circle transiting the meridian is equal to the sidereal time. The date of the year can be found if we can find that date when S.T. = 6 hours and EST = 9:00 p.m. First, as a simplification, we must assume that no longitude correction is possible, and also, we ignore the equation of time. EST = LAT − 12 hours, or LAT = 21 hours. The H.A. of the sun however, is close to 9 hours at 9:00 p.m. (again ignoring

50

the equation of time and any longitude corrections) so, going back to the above relationship:

$$H.A. \text{ (sun)} = S.T. \text{ (6 hours)} - R.A. \text{ (sun)}$$

or

$$R.A. \text{ (sun)} = S.T. \text{ (6 hours)} - H.A. \text{ (sun = 9:00)}$$
$$= -3:00, \text{ or } 24:00 - 3:00 = 21 \text{ hours.}$$

We must now find the date at which the sun's R.A. is 21 hours. Previously, we found that the R.A. of the sun varied during the year from 0 to 24 hours and had the following values on four important dates:

Date	R.A. sun
March 21	0 hours
June 22	6 hours
September 23	12 hours
December 22	18 hours

Since the sun moves 24 hours of R.A. in 12 months, it must travel two hours of R.A. per month. The sun's R.A. on our unknown date is 21 hours. This is three hours greater than its R.A. on December 22, therefore, the proper date is 1½ months after December 22, or about February 10.

Any exercise like the one above can be checked with a rotating star finder. All that is needed is knowledge about the R.A.s and DECs of prominent stars on the chart. These are sometimes listed on the back of various commercial star finders or in elementary textbooks. We will list R.A.s and DECs for several of them:

	R.A.	DEC
Sirius	6½ hrs.	−16½ degrees
Vega	18½	+38½
Capella	5	+46
Arcturus	14	+19
Rigel	5	−08
Procyon	7½	+05
Altair	20	+09
Betelgeuse	6	+07
Aldebaran	4½	+16½
Pollux	7½	+28
Spica	13½	−11
Antares	16½	−26½
Fomalhaut	23	−30
Deneb	20½	+45
Regulus	10	+12
Castor	7½	+32

As another example, when will Betelgeuse rise on March 21? Here, we immediately know the R.A. of the sun (0 hrs.) since on that date it is on the vernal equinox. Also, because Betelgeuse is very close to the celestial equator, its H.A. will be close to 18 hours (or six hours East) when it rises. Its R.A. is about six hours, thus:

$$H.A. = 18 \text{ hours}; \quad R.A. = 6 \text{ hours and}$$
$$S.T. = H.A. + R.A.$$
$$= 18 + 6 = 24 \text{ hours} = 0 \text{ hours}$$

Since the sidereal time is the hour angle of the vernal equinox (R.A. = 0), and on this March date we see the sun at R.A. = 0 hours, the sun's hour angle must also be 24 or 0 hours when Betelgeuse rises. Therefore, the L.A.T. can be calculated:

$$L.A.T. = \text{Hour Angle of Apparent Sun} + 12 \text{ hours}$$
$$= 24 \text{ hours} + 12 \text{ hours} = 36 \text{ hours}.$$

We cannot have time greater than 24 hours, so we must subtract 24 from 36—and find that the L.A.T. = 12 hours, or just at noontime. Betelgeuse will therefore set close to midnight on March 21, and be easily visible in the south at sunset.

It must be understood that this is a simplification of the exact problem. We haven't considered the equation of time or the longitude of observation (to convert L.A.T. to EST), and we have only discussed objects close to the celestial equator. Objects above the celestial equator will rise earlier and set later than those on the equator, and objects south of the celestial equator will rise later and set earlier than comparable objects with similar R.A. but whose declinations are close to zero. Use the sun as a good rule of thumb:

When the sun is at a declination of 23½ degrees north, it rises 1½ hours before an object with the same R.A. but DEC = 0 would, and the sun sets 1½ hours later than this equatorial object. Similarly, when the sun's DEC is 23½ degrees south, it rises 1½ hours later and sets 1½ hours earlier. These approximate values are good only for latitudes close to 40 degrees. Let's use the example of the sun to estimate when a star will set. Antares (R.A. = 16½ hours, DEC = −26½ degrees) is the bright red star in the neck of Scorpius. Its declination is similar to the sun's when the sun is close to the winter solstice. Antares should therefore rise 1½ hours later than an object at the same R.A. but whose DEC is zero.

Question: On May 21, when does Antares set (as seen from Washington, D.C.)?

Answer: First we must determine at what H.A. Antares sets.
Since it sets 1½ hours earlier than its R.A. point on the celestial equator, it must set at 6 − 1½ hours or 4½ hours H.A. Thus:

$$H.A. = S.T. - R.A.$$
$$4½ = S.T. - 16½, \text{ or}$$
$$S.T. = 21 \text{ hours}$$

Now we must determine the L.A.T. equal to S.T. = 21 hours on that date. To do this we must find the sun. What is the sun's R.A. on this date? May 21 is one month before June 21, when the sun's R.A. = 6 hours. Subtracting two hours for the one month time difference, the sun's R.A. = 4 hours on May 21. We can now find the H.A. of the sun and the L.A.T.:

H.A. (sun) = 21 − 4 = 17 hours,

and

L.A.T. = 5 hours or 5 a.m.

So Antares sets near sunrise on May 21. Does this check with your star finder or planetarium demonstration? Antares will be just above the horizon at 5 a.m. EST, according to the finder, on May 21.

Questions

1. When will Pollux rise on April 25? When will it cross the meridian? When will it set?

2. At what latitude will Capella go through the observer's zenith? On May 9, at what time will it pass the meridian as seen from longitude 70 degrees west? Here, you must use the longitude correction to give EST.

3. On May 21, Vega is seen to transit your meridian at 5 a.m. EST. Its meridian altitude was observed to be 65 degrees. What is the latitude and longitude of your observing station?

Observations of the Planets and the Moon

To predict when planets will rise, travel across your meridian, and set, you merely have to find their celestial coordinates, and treat them exactly like stars. The same applies for the moon.

It is not reasonable to expect everyone to carry tables of the positions of planets and the moon around with them all the time. Usually, however, local newspapers and magazines popularizing science announce in which constellations the brighter planets can be found for the month. Once this is known, locate the constellation and you will have found the planet. In most cases it will be the bright object that isn't "twinkling," or the one that doesn't fit the normal configurations of stars in the constellation.

When the moon is in the sky, we usually have no trouble finding it, so we don't have to create any scheme for searching it out. Predicting in advance, however, where and when the moon will be visible is a major problem. Astronomers usually want to avoid having the moon in the sky when they observe faint telescopic objects; the moon is so bright that scattered light from it in our own atmosphere hides most of the rest of the visible universe.

Here is a simple way to predict when the moon will be in the sky, based upon its phases; we will also see how to tell time by the moon from its phases and from where it appears in the sky.

The moon's phases are produced by its varying position in space with respect to the earth and sun. The moon shines by reflected light, of course, and therefore when its illuminated portion faces the earth, it will be visible (see figure 32).

Since the moon is a sphere, only half of its total surface is visible from the earth at any one time. The fraction of that visible half that is illuminated determines its phase. In the scene depicted in figure 32, the earthbound observer would see the earth-facing half of the moon more than 50 percent illuminated. The moon would then be called a "gibbous" moon. If the moon is between the earth and sun, only a thin "crescent" would be observed, as in figure 33.

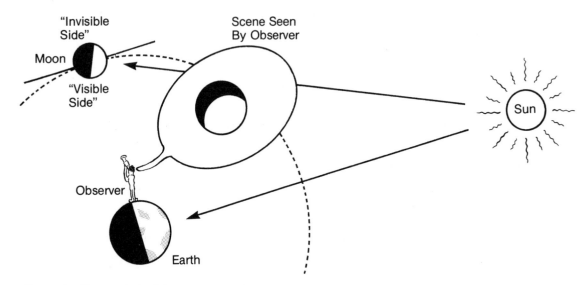

Figure 32 The waxing gibbous moon, as seen from an observer on earth and as seen from a point in space.

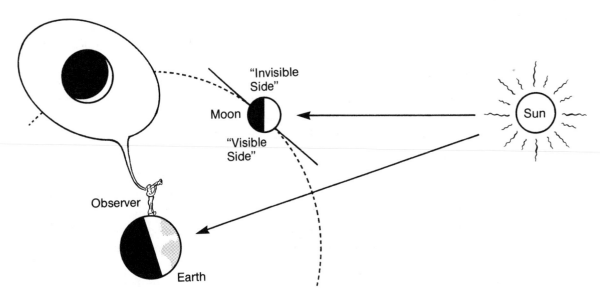

Figure 33 The waxing crescent moon. Only a thin crescent of illuminated lunar surface can be seen from earth.

After staring at the previous examples for awhile, we can draw a general chart that shows the phase of the moon as a function of where it is in relation to the earth and sun (figure 34). Remember, the "invisible" side of the moon always faces away from the earth.

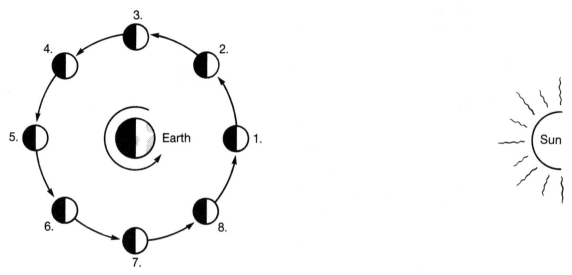

Figure 34 The orbit of the moon showing the moon in various phases. The lighted half of the moon always faces the sun, and different fractions of that lighted half are visible from the earth when the moon is in different parts of its orbit around the earth.

Now let us see what these "phases" look like as seen from the earth, at mid-northern latitudes, and at various times (figure 35).

From these examples, we see that the moon will be found in different parts of the sky relative to the sun when the moon is in its different phases. When the moon is a waxing crescent, it will be seen following the setting sun to the western horizon. When it is a waxing half-illuminated circle (first quarter), it will still be following the sun and will be seen close to the meridian at sunset. Connecting the phase and the position of the moon in space and where it is seen with respect to the sun can be understood by remembering that the earth rotates counterclockwise as seen from its North Pole. The moon's phase progression follows its orbital motion, which also is counterclockwise. Therefore, we can create the following chart:

Position	Phase	Moon is:	(E/W) of Sun
1	New	0 degrees	
2	Waxing Crescent	45	east
3	1st Quarter	90	east
4	Waxing Gibbous	135	east
5	Full	180	east
6	Waning Gibbous	135	west
7	3d Quarter	90	west
8	Waning Crescent	45	west

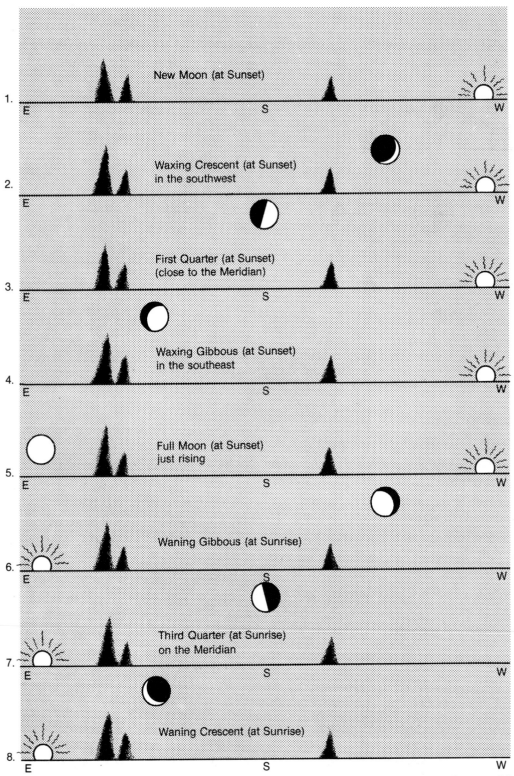

Figure 35 The moon in the sky in different phases, as seen at sunrise and at sunset from a mid-northern latitude.

Remember, start at the sun's position and move CCW (East) or CW (West) until you find the moon. If you wish to know where the moon is in the sky at any time and phase, find the sun and then move east or west by the appropriate number of degrees to find the moon. If the moon is 45 degrees east of the sun, to find the moon you must first find the position of the sun (from the time of day, usually) in the sky, and then move 45 degrees east to find the moon.

If, however, you see the moon in its crescent phase in the sky (and it is waxing—seen in the evening), and want to find the time of day, you must find the position of the sun *from* the moon, and hence you must use the above table backward. Since the waxing crescent is 45 degrees east of the sun, to find the sun you must move 45 degrees west from the observed position of the moon. In this crude way you can find the time of night from the phase of the moon.

The determination of time from the phase of the moon will be accurate (to within an hour) only when the moon is close to the celestial equator. Since the moon's path is really very close to the ecliptic and therefore can be quite distant from the celestial equator, corrections up to $+/-1\frac{1}{2}$ hours might have to be made. In the following two examples, we will assume that the moon is always close to the celestial equator.

Example: The first-quarter moon is rising. What time is it?

Answer: The first-quarter moon is 90 degrees east of the sun. The sun is therefore 90 degrees west of the moon and is leading it by $\frac{1}{4}$ of 24 hours or six hours. Thus if the moon is rising, the sun will already have risen by six hours—it is close to noontime.

Example: The third-quarter moon is rising, what time is it?

Answer: The third-quarter moon is 90 degrees west of the sun. The time will be midnight.

One aid in telling time by the phases of the moon is to recall that the lighted face of the moon always faces the sun. This may seem trivial, but it is easily overlooked. Therefore, if you see a quarter moon in the sky at night, it must be the third-quarter moon if the lighted face is directed east, and the time is probably after midnight.

Corrections to Telling Time by the Moon's Phases:

As with the sun, the moon travels close to the ecliptic. Actually, the lunar path is tilted by 5 degrees to the ecliptic—which for our purposes here is a small enough deviation to ignore. The times of moonrise and set, and the altitude of meridian passage, change throughout the year, as with the sun. To be precise, therefore, we must also consider the season when using this method of time reckoning.

As an example, in the fall, the sun is close to the autumnal equinox. Therefore, the first-quarter moon, 90 degrees east of the sun, is on that part of the ecliptic that is normally occupied by the sun three months hence, or in winter, and so the moon has a declination of around − 23 degrees. Its time of rise is approximately 1½ hours later than the rise time of its R.A. point on the celestial equator. Its set time is 1½ hours earlier than the set time of its R.A. point of the celestial equator (see figure 36).

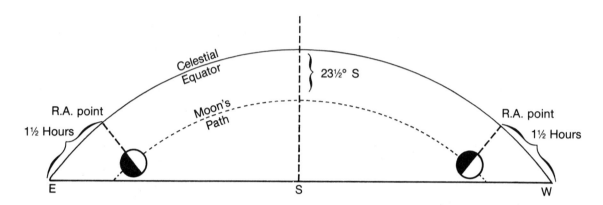

Figure 36 The path of the moon as seen looking south when the moon is at first quarter during the fall months.

Let us look at other lunar phases during the fall. The full moon is 180 degrees east of sun and is therefore sitting close to the vernal equinox and hence is on the celestial equator. It will rise with its R.A. point, spend about 12 hours in the sky, and will set with its R.A. point. No corrections would therefore be needed.

The third-quarter moon, which is 90 degrees west of the sun, is also at the summer solstice point of the ecliptic at declination + 23 degrees north. It rises 1½ hours earlier than its R.A. point, and sets 1½ hours later than its R.A. point. It behaves like the summer sun.

Turning now to the winter, the sun is close to the winter solstice and has a declination of − 23 degrees south. So the first-quarter moon will be at the vernal equinox and therefore will be close to the celestial equator. No corrections would be needed.

In the winter, the full moon will be close to the summer solstice point—so treat it as a summer sun. The third-quarter moon will be close to the autumnal equinox. Therefore no corrections are needed.

In the spring, the sun is on the celestial equator close to the vernal equinox. The first-quarter moon will be at the summer solstice point, so it should be treated as a summer sun. The full moon is close to the autumnal equinox, so no corrections are needed. The third-quarter moon is close to the winter solstice, so it should be treated as a winter sun.

In the summer, with the sun close to the summer solstice, the first-quarter moon will be at the autumnal equinox, so no corrections are needed. The full moon will be at the winter solstice, so it should be treated as a winter sun, and the third-quarter moon is at the vernal equinox and so no corrections are needed.

We see from the above that we merely have to find out which part of the ecliptic is occupied by the moon at its particular phase, and treat its rise and set times as we would the sun's corrections to rise and set on those dates.

Example: In the spring, at what time does the first-quarter moon set?

Answer: In the spring, the first-quarter moon is in the summer sun's position. It therefore sets 1½ hours later than midnight, or around 1:30 a.m. The last-quarter moon, sitting at the winter solstice position, will rise at 1:30 a.m.

Questions

(All assume mid-northern latitudes, unless otherwise stated.)

1. The date is fall and the third quarter moon is rising, what time is it?

2. What is the meridian altitude of the full moon in the winter? In the summer? How does this affect nighttime illumination?

3. What time of year will the first quarter moon be visible in the sky for 12 hours?

4. The waxing crescent moon is low in the southwest, what time of night is it if the season is fall?

Hint: For all phases of the moon other than new, quarter or full, determine between which two phases the phase in questions falls, and average your correction. For example: The waxing crescent is between new moon and first quarter moon. In the fall, its correction would be half of that for the first quarter or three-quarters of an hour earlier for set time and three-quarters of an hour later for rise time.

5. The waning gibbous moon is on your meridian. What time is it if the season is spring?

6. From the Arctic Circle, on what date will the full moon be above the horizon for 24 hours?

Planetarium Exercise   (data sheet, p. 105)

Printed here is a data sheet to be filled in during a planetarium demonstration of the phases of the moon. This data sheet should then act as a general guide for telling time by the moon's phases. As in the hint above, for questions involving phases other than the standard ones at quarter and full, simply interpolate to find the average correction (always $+/- $ ¾ hour).

From the data sheet, you should be able to draw up a chart listing times, phases of the moon, and you might add their horizon positions in the sky (E, SE, S, SW, W, NE, NW, etc.).

5   Eclipses of the Sun and Moon

In the last chapter we examined the various phases of the moon and became familiar with its motions. We know that once every lunar month, the moon travels between the earth and sun and then, a half lunar month later, the earth comes between the moon and sun. When these conditions are exactly satisfied, and the earth, moon, and sun lie along one line in space, eclipses will occur. If the moon's orbit were exactly aligned with the plane of the ecliptic, we would have lunar eclipses and solar eclipses every month, as seen from some place on earth. But this does not happen. Lunar eclipses occur infrequently and at seemingly irregular intervals, and solar eclipses, especially total solar eclipses, are very rarely seen from any one place on earth.

There are a number of reasons for the comparative rarity of eclipses. Let us first look at the conditions needed for an eclipse to occur, then we might better understand why they are infrequent events.

For any kind of eclipse to occur, one body must pass into the shadow of another body (see figure 37). The sun, illuminating both bodies, will then be eclipsed as seen from the body in the shadow. A person sitting on the body producing the shadow will see nothing irregular, except that the second body within the shadow will no longer shine by light reflected from the sun.

A lunar eclipse occurs when the moon passes into the earth's shadow. Since the moon has to be exactly opposite the sun as seen from the earth, lunar eclipses must therefore occur sometime around full moon. The earth's shadow, extended out into space, is still considerably larger than the disk of the moon at the point where it crosses the lunar orbit. Therefore, the moon does not have to be exactly on the ecliptic for a lunar eclipse to occur.

A solar eclipse takes place when the moon passes between the earth and the sun, and when the moon's shadow reaches the surface of the earth. At the moon's mean distance from the earth, its shadow just barely reaches the earth. As with all celestial orbits (of planets, moons, comets, etc.), the moon's orbit is not circular; if you looked at the orbital path carefully, and measured several diameters, you would find that it was slightly squashed into an elliptical shape. The ellipticity of the orbit is such that when the moon is at distant parts of its orbit around the earth, if all other conditions for an eclipse were to occur, there would still be no total eclipse observed. When the moon is farthest from the earth, it is at a distance greater than the length of its own shadow, and the eclipse becomes annular. The orientation of the orbital ellipse of the moon does not remain in a constant position with respect to the sun.

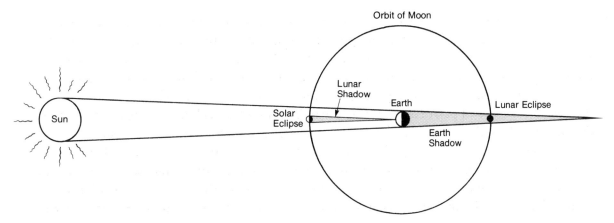

Figure 37 Configuration required for solar and lunar eclipses, as seen perpendicular to the orbit of the earth.

From the above, it should be clear now why eclipses of the sun are rare. Even when the moon is at its closest distance to the earth, and all other conditions for an eclipse of the sun are satisfied, the size of the lunar shadow reaching the earth's surface is very small, usually less than 100 miles in width and normally many miles less than that. Therefore, the region from which a total solar eclipse might be visible is extremely small, and your chance of seeing a total solar eclipse without traveling to some other site is accordingly very small. For this reason, it is not surprising that some enterprising travel agencies arrange and conduct guided tours that cross predicted eclipse paths. The most popular seem to be ocean cruises, since ocean liners are free to move to the spots most likely to have clear weather. Ocean cruises seem to be popular for other reasons too, but they are not astronomical!

As we mentioned before, the moon's orbit is not exactly in the plane of the ecliptic; it is tilted by about 5 degrees. If we examine figure 38, which represents the worst possible conditions for an eclipse, we see that the moon misses the earth's shadow, and the lunar shadow misses the earth. In this figure, the intersection of the moon's orbit and the earth's

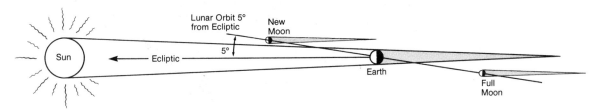

Figure 38 The earth-moon-sun system as seen in the plane of the earth's orbit. The line of nodes of the earth's orbit and the moon's orbit point out of the page. In this configuration, eclipses do not occur.

orbit is a line perpendicular to the page, going through the earth. If we were to rotate the orbit of the moon through 90 degrees, so that this line of intersection of the two planes points toward the sun, then eclipses will occur, as figure 39 shows in perspective. In this figure, the orbit of the earth and the orbit of the moon intersect in a line passing through the earth, moon and sun. This satisfies all conditions (except the distance to the moon) for a solar eclipse to be total.

A more functional way to look at the situation is to take the projection titled "plane of sky" in figure 39 and expand it as in figure 40. In this projection, we see that a total solar eclipse can occur when the sun and moon are both in the shaded areas of their orbits. Figure 41 represents the situation better, since the 5 degree tilt is quite small. In figure 41, if the sun is anywhere within the shaded area during the course of a lunar month when the moon passes through, some type of solar eclipse (total, partial, or annular) will occur. The period of time the sun spends within this shaded area is called, appropriately, an eclipse season.

We have drawn the two paths as if both the sun and the moon subtended angular diameters of half a degree in our sky, which is quite close to the actual value. What we have not included here is the fact that the effective angular diameter of the sun should be a bit bigger than its actual angular diameter, because eclipses can occur anywhere on the face of the earth,

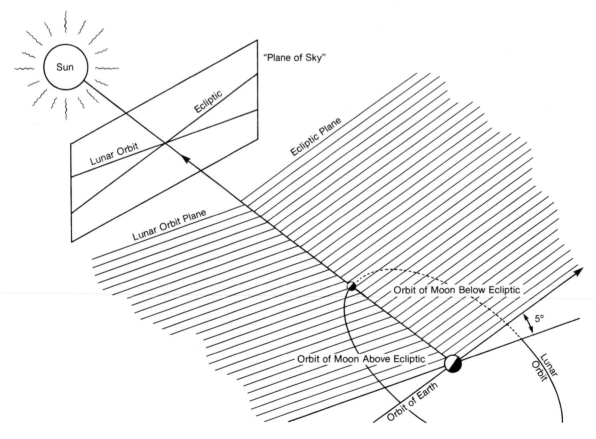

Figure 39 Oblique view of the earth-moon-sun system with the intersection of the earth's orbit and the lunar orbit lying in the direction of the sun—a condition necessary for eclipses to occur.

which subtends a full 2 degrees in the lunar sky. Thus, if the moon passed just a bit north of the sun, its shadow would hit somewhere on the northern surface of the earth. This fact relaxes the requirements for solar eclipses, as well as lunar eclipses.

Requirements for lunar eclipse visibility are less stringent, however, since everyone on the nighttime half of the earth's surface at the time of the eclipse will see it. This is the major reason why lunar eclipses are more frequently observed than solar eclipses.

The durations of lunar and solar eclipses are quite different. Neglecting the motion of the earth around the sun (which causes the apparent motion of the sun on the ecliptic), we can examine the lengths of solar and lunar eclipses by noting how quickly the moon can move across the solar disk or how quickly it can pass through the earth's shadow.

We will assume central eclipses in both cases, the angular diameter of the sun and moon to be both ½ degree, and the angular diameter of the earth to be 2 degrees, as seen from the moon. In one lunar month of approximately 27 days, the moon "moves" 360 degrees in the sky, as seen from the earth. In one day, therefore, the moon will move about 13.3 degrees in its orbit, and thus it moves approximately ½ degree per hour.

The time between "first contact" in a solar eclipse (when the leading limb of the moon first touches the trailing limb of the sun) and "last contact" (when the trailing limb of the

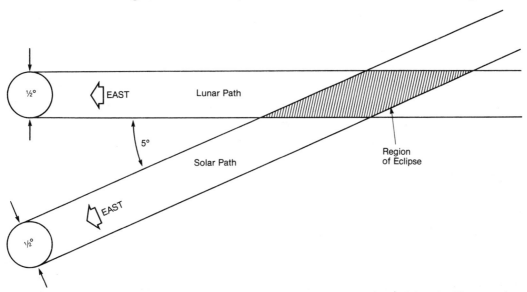

Figure 40 The intersection of the lunar and solar paths, as seen in the plane of the sky. The crosshatched region identifies that part of the two paths where an eclipse might occur. Here the inclination of the lunar orbit is greatly exaggerated.

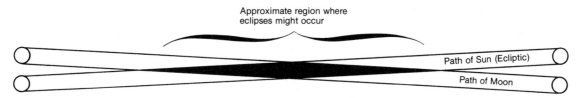

Figure 41 The actual inclination of the lunar orbit with respect to the earth's orbit around the sun—only five degrees—creates a long eclipse season.

moon leaves the leading limb of the sun) must then be the amount of time it takes the moon to move its diameter plus the solar diameter: ½ + ½ = 1 degree, or two hours. Most of this will be the partial phase of the solar eclipse, of course; the period of totality being much less, never more than a few minutes.

The time between first and last contact during a total lunar eclipse is somewhat greater, since the moon must move an equivalent of its own diameter (½ degree) plus the angular diameter of the earth as seen from the moon (2 degrees) for a total of 2½ degrees. Moving at ½ degree per hour, a complete eclipse can take five hours, but totality is shorter. For totality, the moon must be completely immersed in the earth's shadow. Totality starts when the center of the moon is ¼ degree within the shadow, and ends when the center of the moon approaches to within ¼ degree of the trailing edge of the shadow. This reduces maximum totality to three hours.

Lunar eclipses occur quite slowly, and can be observed at ease. Partial lunar eclipses, and "penumbral" eclipses of the moon, occur even more frequently, but are not very interesting events, being detectable only after very close examination.

Today we consider lunar and solar eclipses to be interesting astronomical phenomena, and solar eclipses are still important events for science. Only during a total solar eclipse is the tenuous outer atmosphere of the sun visible for study from the earth, when the vastly brighter solar disk is cut from our view by the disk of the moon. Aside from its scientific interest, a solar eclipse is possibly the most awe-inspiring astronomical event one can normally experience. In an instant, day turns to night, with a reddish-golden horizon of twilight spanning 360 degrees. Birds hover or roost, stars become visible, and the temperature drops noticably. As the sun disappears, supposedly without warning, it is not difficult to understand why ancient civilizations regarded these events with fear and trepidation.

In later chapters of this workbook, we will see how solar and lunar eclipses were used by the Greeks to provide us with our first rough ideas of the scale of the solar system, and hence of the visible universe.

Questions

1. If the orbit of the moon was coincident with the ecliptic, approximately how many solar eclipses would be visible from earth each year? Would they all be visible from any part of the earth?

2. Why have more people seen a lunar eclipse than a solar eclipse?

3. Name a famous story from American literature that involves a solar eclipse. How was the eclipse represented?

4. If the moon were half as far away, how would this change the number of solar eclipses seen? How would this change the number of lunar eclipses seen? Would there be any effect on their duration?

6 Ancient Cosmological Speculations

The first part of our study has shown how the sun, moon, and planets seem to move—as seen from a moving earth. Only in recent times has mankind generally accepted the motions of the earth (its revolution around the sun and its rotation about its own axis). To the ancients, the earth was stationary, and all celestial bodies traveled around it.

Before the Greeks, the motions of the sun, moon and planets were cataloged year in and year out; calendars and counting systems were produced that would allow for the empirical predictions of future positions of the celestial bodies. The Greeks, however, created a geometrical model that could predict planetary, solar, and lunar positions. They had to account, therefore, for the following irregularities in celestial motion:

1. The seasons—and their unequal lengths.

2. The unequal amounts of time needed for the moon to progress from one phase to another; the time from first-quarter moon to third-quarter moon was not the same as the time from third to first quarter.

3. The varying brightnesses of the planets and their irregular motions.

The third irregularity was the most perplexing; even though irregular motions were accounted for reasonably (as we shall see), the varying brightness of the planets was never adequately accounted for.

The Greeks wanted to create a model of the solar system that would describe the motions of the sun, moon and planets. No special reality was given to the mechanism that was created to do this. The character of the mechanism that was required was of intellectual interest; the Greeks hoped they could find some way to describe how the sun, moon, and planets moved about the earth. They also hoped for a convenient model that could be used to satisfy the social need of predicting celestial positions and hence important seasons, calendar events, celebrations, and tax dates.

The geometric mechanism used by the Greeks to represent celestial motion was the circle. All celestial objects moved on circles or spheres, or combinations of circles, and all motion was uniform. With these constraints, and the belief that the earth was stationary at the center of all motion, the phenomena mentioned above became difficult to describe.

Retrograde Motions of the Planets

Since prehistory, planets were observed not to travel in even, uniform paths on the celestial sphere. The two **inferior** (closer to the sun than the earth) planets, Mercury and Venus, always traveled close to the sun, and in fact seemed to circle its position in the sky. While the sun traveled more or less at an even eastward rate among the stars, Mercury and Venus only sometimes traveled easterly. At various times, their regular eastward motions would stop, and they would travel westward for a while, generally circling the sun (see figure 42).

The **superior** (farther from the sun than the earth) planets appeared to behave in much the same manner. Most of their motion was eastward, but every so often, this eastward motion would stop, and the planets would back up, or travel westward. This westward motion, called **retrograde motion,** is depicted over the passage of time on the celestial sphere in figure 43.

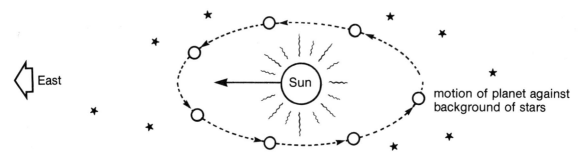

Figure 42 Retrograde loop for an inferior planet, as seen from earth over several weeks' time.

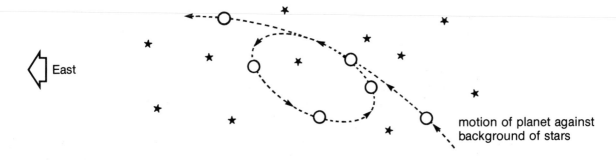

Figure 43 Retrograde loop for a superior planet, as seen in the plane of the sky over several weeks' time.

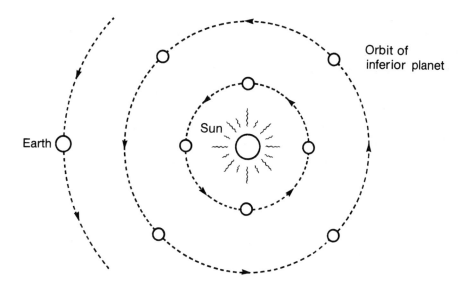

Figure 44 The orbits of two inferior planets, as seen from a point in space directly above the plane of the earth's orbit.

From a moving earth, this apparent motion is not difficult to understand. Inferior planets (Mercury and Venus) travel about the sun in orbits smaller than that of the earth, and therefore seem to circle the sun. The extra-solar view in figure 44 should be compared to the terrestrial view in figure 42.

Superior planets (Mars, Jupiter, Saturn) travel slower in their orbits around the sun than does the earth. The earth, therefore, can pass these planets in their orbits from time to time, and when it does, the superior planets apparently travel backward, as seen in figure 45.

Our modern heliocentric view of the structure of the solar system (sun at center with planets traveling about it) appears to explain the retrograde motions of superior planets. The Greeks, however, preferred to attempt descriptions and explanations based upon a stationary earth, allowing only uniform, circular motion to occur. The varying motion of the sun was produced by a circular orbit slightly displaced from the earth's position, as seen in figure 46.

Accounting for the retrograde motions of the superior planets, however, was not so simple a task. Here, some additional circular motion had to be superimposed on the basic circular motion This was accomplished by introducing "epicyclic motion"—the planet now was placed upon a small circle that rotated around a larger circle, which in turn circled a point close to earth (see figure 47). The small circle is called an **epicycle** and the large circle a **deferent.**

The small epicycle motion could modify the effect of the uniform circular motion of the deferent. In figure 48, an epicycle with a rotation period equal to that of the deferent is seen in opposite parts of its orbit. On the right, its motion counters that of the deferent, while on the left, the motions enhance one another. Note: an epicycle with no rotation of its own will still go through one rotation as seen from the outside, since the deferent it is riding on makes one revolution.

If the motion of the epicycle and deferent were correctly chosen, a planet could be made to travel backward in its orbit, as seen from a stationary central earth. All the while, of course, the planet would be moving with constant uniform circular motion on its epicycle and the epicycle would be moving constantly around the deferent. Thus, uniform circular motion could be combined into what appears to be non-uniform motion.

One of the most useful generalizations of epicyclic motion was made by the Greek astronomer Hipparchus (approximately 150 B.C.E.). He was able to show that epicyclic motion was equivalent to displaced (or eccentric) circular motion. We will provide a modern example that, while not historically precise, still conveys the character of what he did.

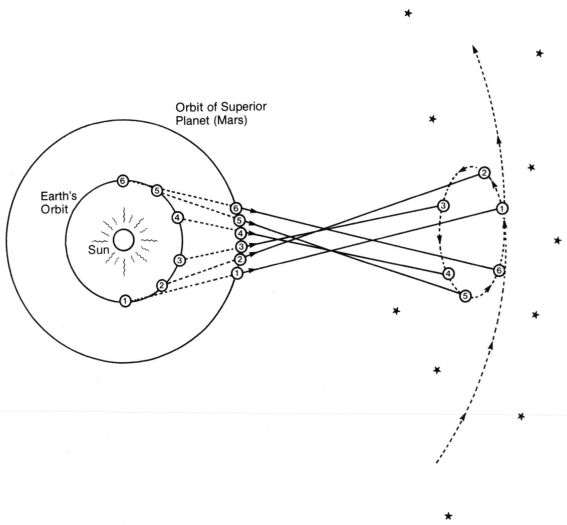

Figure 45 The orbital paths of the earth and a superior planet, with both planets depicted at six points in time. Note that the earth, on its inner orbit, passes the superior planet, causing it to apparently move backward against the background of stars. The line-of-sight arrows from the earth through the planet's position are then projected against the plane of the sky, showing that the object retrogrades as the earth passes the superior planet.

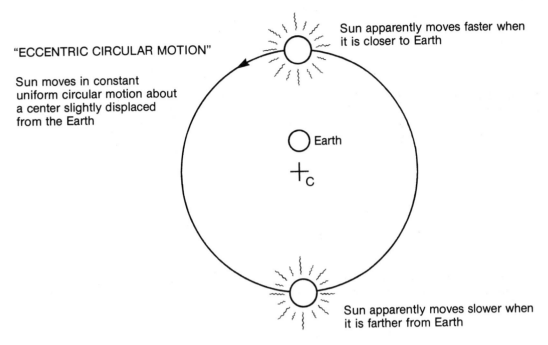

"ECCENTRIC CIRCULAR MOTION"

Sun moves in constant
uniform circular motion about
a center slightly displaced
from the Earth

Sun apparently moves faster when
it is closer to Earth

Earth

+ C

Sun apparently moves slower when
it is farther from Earth

Figure 46 Eccentric circular motion. The motion of the sun about a point displaced from the earth provides seasonally uneven rates of apparent solar motion. The sun will apparently move slower when it is more distant from the earth. All the while, the sun is executing constant circular motion about the center of the deferential circle.

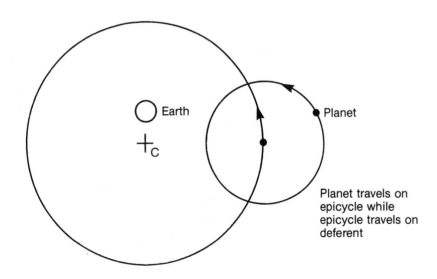

Earth

+ C

Planet

Planet travels on
epicycle while
epicycle travels on
deferent

Figure 47 Epicyclic motion. Planet P travels on a small circle called an epicycle, whose center in turn is carried around the earth by the motion of a larger circle, called the deferent. Again, the earth is displaced from the center of the deferent.

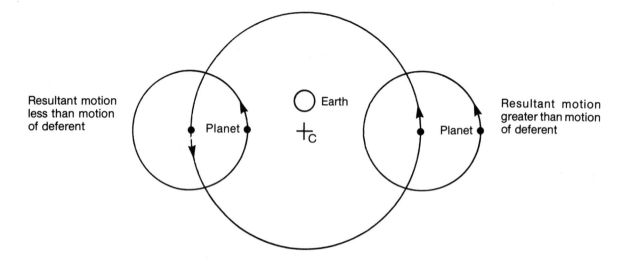

Figure 48 The resultant motion of planet P on its epicycle and deferent when the epicycle aids in the direction of motion of the deferent and when it cancels the motion of the deferent.

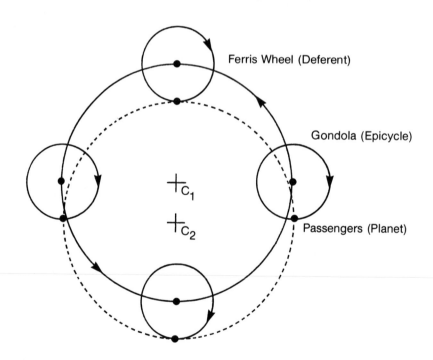

Figure 49 The motions of passengers on a Ferris wheel are similar to those of a planet moving on an epicycle that itself is moving on a deferent around the earth. The period of each rotation of the epicycles is equal to the period of rotation of the deferent. The resultant motion of each passenger is therefore circular, but displaced from the center of the deferent (or main wheel of the Ferris wheel). C1 is the center of the Ferris wheel, C2 is the center of the effective motion of each passenger.

Imagine a large Ferris wheel with cylindrical gondolas as carriages that are free to move on the main circle of the Ferris wheel. The rotation of the gondolas would always permit the passengers to be sitting properly as the Ferris wheel rotated.

If the passengers inside the gondolas represent the motion of a planet around the earth, which is assumed to be near the center of the Ferris wheel, then the rotation of the gondolas (or epicycles) would have to be equal to the rotation rate of the main Ferris wheel (or the deferent). The passengers would still be moving in a circle, but their motions would now be displaced from the center of the Ferris wheel by an amount equal to the radius of each gondola. So, displaced circular motion is equivalent to epicyclic motion.

In figure 49, C1 is the center of the Ferris wheel (or deferent) while C2 is the center of the resultant motion of the passengers in the gondolas (or planets on the epicycles). For this to work, the Ferris wheel has to rotate counterclockwise while the gondolas have to rotate clockwise.

Since the effects of epicyclic motion and eccentric circular motion were equivalent in Greek modeling, they could be used in combination to account for all sorts of irregular motions. This generalized combination of motions was eventually, in the hands of Claudius Ptolemy (fl 140 A.D.), able to account quite well for the motions of the sun, moon, and planets. Ptolemy reviewed all of the earlier Greek accomplishments in planetary theory in his 13-volume *Almagest,* and through it formalized what we today call the Greek Geocentric Universe.

7 The Dimensions of Space

Hipparchus is regarded as the greatest of all Greek astronomers. In addition to his study of epicyclic motion, he conducted systematic observations of the positions of the planets, sun and moon against the background of stars. His observations were highly accurate and formed the basis for Ptolemy's catalogue of star and planet positions.

Hipparchus and Greek mathematicians of his time also created the mathematical short-hand known as trigonometry. With this shorthand, they were able to say what the relative sizes of the sides of triangles were simply from knowing their various angles. This was a tremendous advance in the art of calculating distances. Triangulation became a quick, accurate, and efficient method of determining distances and heights of things.

Earlier, if someone wanted to know the height of a tree or the distance across a river, baselines would have to be measured and angles observed, and then the resultant triangular relationship would have to be tediously drawn out on paper. But Hipparchus, along with Greek and Oriental mathematicians, tabulated the sizes of the various sides of triangles in terms of these angles, generating them either algebraically or mechanically. Thus, when a new observation was made, only the table had to be consulted.

A number of simple definitions were created that equated the sizes of angles within special right-angled triangles to the sizes of the sides of these triangles (see figure 50).

In a later section we tabulate values of the **sine** in our discussion of how Copernicus calculated the relative orbital sizes of the planetary orbits. But now we turn from Greek cosmology to study how the Greeks utilized triangulation in order to determine the scale of the solar system.

To appreciate the Greek technique of indirect measurement, we will determine the width of a river. Figure 51 shows how, by sighting on an object on the far side of a river from two points—one directly opposite so that the angle between the bank of the river and the object on the opposite side is 90 degrees and the other from a point down the bank, an easily measurable distance (say 100 feet)—a triangle is created that can yield the width of the river. Study the

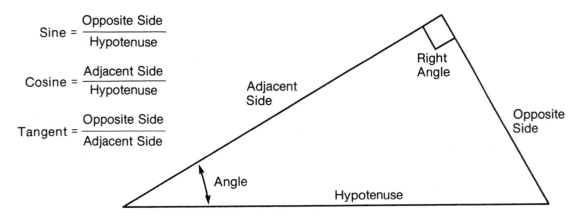

Figure 50 The relationships between the lengths of the sides and the sizes of the angles in a right triangle.

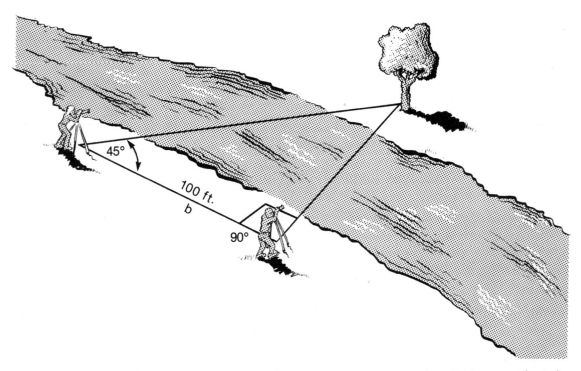

Figure 51 Determining the width of a river by triangulation. Two sighting points 100 feet apart form the base of the triangle, and the spot where the line of sights from both points converge represents the object sighted on the opposite river bank.

triangle and determine which function is needed to determine the width of the river. If the baseline along the river bank (b) is 100 feet and the angle is 45 degrees, the width of the river is also 100 feet.

We have now had some experience with the method of triangulation, but why the fuss? Well, it just so happens that the only methods available to measure the distances to planets, the distance to the moon, and the radius of the earth required geometry and rudimentary triangulation before the twentieth century.

Now that we have looked at the basics, I hope that you will be able to apply them to the more complex ideas and techniques that follow. Remember, however, that whatever the reasons for making these determinations in the past, we are reviewing them here to see how we came to know the characteristics of the solar system that are of fundamental importance in deriving the scale of the visible universe and the fundamental laws of dynamics that drive it. Note too that even though we opened this chapter reviewing the mathematical shorthand of trigonometry that eventually will yield the dimensions of the solar system, the first steps were taken without trigonometry. Also, rather than presenting the methods of determining the dimensions of space in chronological fashion, we proceed from the earth's dimensions out into space.

The Radius of the Earth

Eratosthenes (250 B.C.E.) is regarded as the first person to determine successfully the radius of the earth. He was a geographer and so was aware of the need to calibrate his maps in terms of the actual size of the earth. Eratosthenes possessed a great amount of astronomical data because of his association with the library at Alexandria. From records kept there he knew that on the longest day of the year (June 22) the bottom of a deep well in the town of Syene was illuminated so the sun was directly overhead at that place on that date.

Question: From this observation, what was the latitude of Syene? Answer: 23½ degrees north.

As the story goes, on that date Eratosthenes measured the altitude of the sun at noon from Alexandria (which was fortunately due north of Syene—why is this important?) and by using a little common sense was able to deduce the size of the earth. Figure 52 illustrates the geometry of the situation. A triangle can be drawn between the towns of Alexandria and Syene and the center of the earth. Once the angle subtended by the two towns, as seen from the center of the earth, is determined from the angle of the sun's shadow at Alexandria, the circumference of the earth can be found.

The length of the shadow cast by the gnomon at Alexandria, compared to the height of the gnomon, indicated that the angle of the deviation of the sun from the vertical gnomon was very small, some ⅟₅₀th of a circle. We have designated this angle as Theta (Θ).

Assuming that the rays of sunlight at Alexandria and at Syene are parallel, the radius vector through Alexandria cuts the sun's rays as a transversal between two parallel lines. The two alternate interior angles thereby produced are equal, and so the distance from Alexandria to Syene is also ⅟₅₀th of the distance around the earth. This distance was known to be 5,000 "stadia," or about 500 miles in today's unit of distance. So the circumference of the earth was found to be 50 × 500 miles, or 25,000 miles, and the earth's radius, 25,000/2pi = 4,000

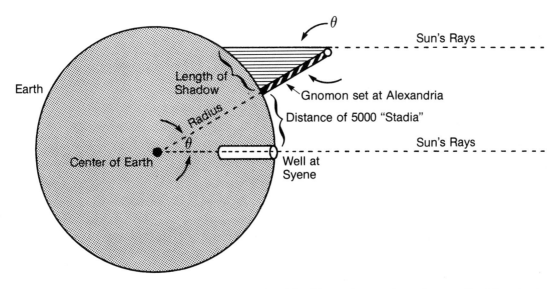

Figure 52 The radius of the earth as determined by Eratosthenes, based upon the triangle drawn between Syene, Alexandria, and the center of the earth.

(where pi = 3.14).

In modern terms this was a very accurate determination. In fact, however, there is only very weak (if any) real evidence showing that Eratosthenes actually made this measurement or performed exacting observations and calculations. Quite possibly, he developed it as a "thought" experiment, knowing from various records what the altitude of the sun was from those two places on that date. Historians also are far from sure just how large "stadia" were.

Question: The sun's rays are not parallel but divergent. How does this affect the diameter of the earth? Does it make the earth larger or smaller?

Answer: Eratosthenes believed that the sun was so far away that its position in space would hardly shift in space as seen from two points on opposite sides of the earth. He got this idea from the earlier work of Aristarchus (275 B.C.E.), who determined that the sun was at least 20 times farther away from the earth than was the moon. The slightly divergent rays of the sun, not accounted for, make the shadow angle at Alexandria a bit larger than if the rays were parallel. A larger angle meant that the distance between Alexandria and Syene was a larger fraction of the circumference of the earth, thus making the earth appear smaller than it was.

Distances to the Sun and Moon

Aristarchus noticed that when the line between day and night on the visible surface of the moon (the moon's terminater) was straight, the moon itself was not exactly at right angles with respect to the sun's position in the sky (see figure 53).

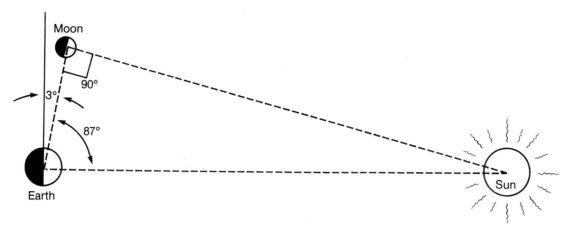

Figure 53 The geometry Aristarchus employed to determine that the sun was 20 times farther away from the earth than was the moon.

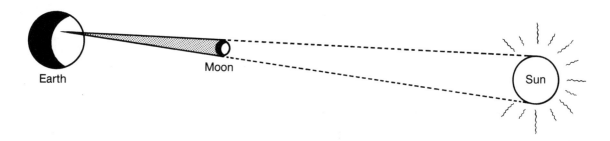

Figure 54 As seen from any point on the earth, the sun and moon are sensibly of the same angular size. Otherwise, eclipses of the sun would not happen in the way they are seen.

Knowing that the moon is illuminated by the sun, Aristarchus devised a method—based upon his observation of the position of the moon when the terminater was straight—to determine the relative distances to the moon and sun. When the terminator was straight, the angle between the sun, moon, and earth must be a right angle. By determining somehow the angle between the moon, earth, and sun, Aristarchus produced a skinny triangle with the earth-moon distance as one side and the earth-sun distance the hypoteneuse. In figure 53, the moon-earth-sun angle is A and the sun-moon-earth angle is C = 90 degrees.

By various techniques, Aristarchus found a value that in today's mathematical language would be about 87 degrees. Therefore, EM/ES = cosine (87 degrees) = 1/20th of a circle, although Aristarchus did not have **cosines** to use. In any event, from this simple exercise, we might appreciate how Aristarchus determined that the moon was 20 times closer to the earth than was the sun.

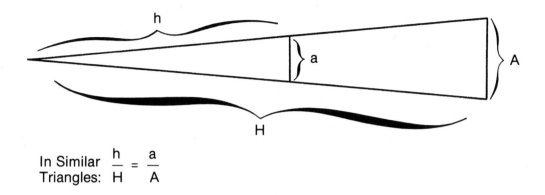

In Similar
Triangles: $\dfrac{h}{H} = \dfrac{a}{A}$

Figure 55 The similar triangle relationship described by the relative sizes of the earth, sun, and moon, as seen in figure 56.

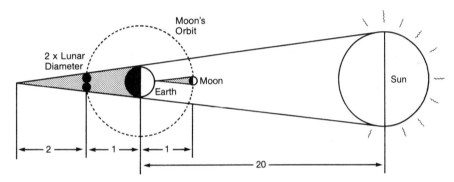

Figure 56 The triangle defined by the shadow cone of the earth creates three similar triangles: the largest has the sun as its base, the next largest has the earth as its base, and the smallest has twice the moon's diameter as its base.

Relative Sizes of the Earth, Moon, and Sun

From continued observations of solar eclipses Aristarchus also knew that the angular sizes of the sun and moon were equal, as in figure 54. Otherwise eclipses would not occur in the way they are seen to happen. If the distance to the sun was known, its size relative to the moon could be determined. The geometry of similar triangles was all that was needed to make the determination.

Figure 55 shows the two basic triangles involved; one is nested within the other. The sun was found to be 20 times farther away than the moon and so it had to be about 20 times larger than the moon. This made the sun about five times larger than the earth.

The idea behind Aristarchus' method of similar triangles is presented in figure 55. Here, triangles AH and ah are similar in that their two sides coincide and the third sides (the bases) are parallel. Therefore, the ratio of the height to the base of each triangle is constant. If the triangle represents the earth-sun system, it would look like figure 56.

The triangles involved have bases of the solar diameter, and the total height of the triangular cone as one triangle; the base of the earth's diameter and the height of the triangle cone above the earth as a second triangle; and the base of two lunar diameters (determined by the timed length of lunar eclipses) with the height of two lunar distances as the third triangle.

Aristarchus determined the length of the earth's shadow beyond the moon by noting that the moon's shadow just barely reaches the earth, and is hence one lunar distance in length. So the length of the shadow cone at a point where its base is twice the lunar diameter would be twice the lunar-shadow distance. The total triangle is thus 23 lunar units in height; the triangle above the earth is three lunar units, and the triangle above twice the moon's diameter is two lunar distance units. Thus the ratios of heights to bases become:

$$\frac{\text{Solar diameter}}{23} = \frac{\text{Earth's diameter}}{3} = \frac{2 \times \text{Moon's diameter}}{2}$$

So to Aristarchus, the sun is 23 times the size of the moon, and is about seven to eight times the size of the earth.

The great size of the sun so calculated stimulated Aristarchus to think that the earth was not the center of the universe. The sun, so much larger than the earth, and also the source of light and heat, certainly had to be the center of the universe.

If the earth indeed circled the sun, however, the Greeks believed that two things should happen. First, stars on the celestial sphere would shift relative to one another due to the shifting position of the earth around the sun. This was Aristotle's argument. It was later revived by Tycho Brahe. Second, if the earth was in motion, then we should surely feel it. Aristotle again suggested this argument.

Aristotle shouldn't be scolded for a few ideas that later proved to be all wet; for over a thousand years, everyone believed him without question. Neither of these problems were even close to being resolved when Copernicus's revolutionary ideas were published in 1543. Although there were pressures on Copernicus to reveal the results of his calculations and convictions, when they did appear (after his death) no great revolution occurred immediately. More than a hundred and fifty years passed before a combination of ideas, observations, and techniques appeared that permitted an explanation of the movement of the earth and the other planets and provided a framework for further scientific development.

With this in mind we next examine the works of Copernicus, Tycho, Kepler, Galileo, and finally Newton. Remember, it took the greatest minds almost two hundred years after Copernicus to reorient the earth and the fundamental nature of the physical universe. We can only devote a few hours to it.

Copernicus

Nicholas Copernicus (1473–1543) was trained in the law and medicine, but became completely absorbed by the study of astronomy. He examined revived texts and translations from Greece that had passed through the Moslem world into the European world during the past 1,400 years, and developed a strong conviction that a very different astronomical view was necessary: one in which the earth moved around the sun. His new world system, which first appeared in his writings around 1512, had seven major points:

1. There is no single center of motion in the universe.

2. The earth is not the center of the universe; it is only the center of "gravity" for the earth and moon.

3. Everything except the moon circles the sun.

4. Stellar distances are "unknowable."

5. The diurnal motions of stars are caused by the motion of the earth on its axis, its rotation.

6. The daily and annual motion of the sun is due to motions of the earth.

7. The retrograde motions of the planets are due to the relative motions of the planets and earth.

We have already seen how the Copernican system explains retrograde motion. Copernicus also applied his model to the determination of the relative sizes of the planetary orbits in the solar system.

The Relative Distances to the Planets from the Sun (data sheet, p. 106)

The planetarium offers us the ability to move rapidly in time. The distances of the various planets from the sun (or the sizes of their orbits) can be determined by plotting their changing positions among the stars with respect to the sun over the course of many months. Copernicus was the first to do this although he had to wait for very long periods of time while the planets moved in their courses or he depended upon the collected data on planetary motions produced by earlier astronomers. In a planetarium we can accomplish the needed observations quite quickly.

Mercury and Venus

We will examine Venus, knowing that the technique is identical for Mercury. First of all, Venus is an *inferior* planet with respect to the earth. It therefore never gets very far from the sun in the sky, as seen from the earth. Indeed, Venus has always been referred to as the "Morning" or "Evening" star. If we travel out into space and look back upon the solar system we find the reason, as depicted in figure 57.

The dotted lines in figure 57 represent the maximum angular distance that Venus can appear to be separated from the sun, as seen from the earth. Labeling those points where the dotted lines cross the Venusian orbit as T, we can see that a simple relationship exists between the size of the orbit (TS) and the maximum angular elongation of Venus. In figure 58, the Venusian orbit (TS) = the earth's orbital size (ES) × sine (Θ).

From figure 58, we can see that:

$$\text{sine } (\Theta) = \text{TS/ES, or}$$
$$\text{Venusian Orbit (TS)} = \text{Earth Orbit (ES)} \times \text{sine } (\Theta)$$

In the planetarium we will measure the angular distance of Venus from the sun on successive dates. This will reveal its maximum angle of elongation (Θ).

On a separate piece of paper (based on the sample data sheet), we will plot Θ against the date of the observation and find the maximum value of Θ, from which we will derive the orbital radius of Venus. An example is provided in figure 59. Note in the figure that a smooth curve through the observations reveals a point, not necessarily at a point of observation, where Venus is at a maximum angular distance from the sun, as seen from earth.

The accompanying table of sines will allow you to convert the maximum angular value into an orbital dimension. If you find a maximum angle of elongation of 34 degrees, for example, then:

$$\text{Venusian orbit (TS)} = \text{earth's orbit (ES)} \times \text{sine } (34),$$
using the table, the sine of 34 degrees is 0.56, so:
$$\text{Venusian orbit} = 0.56 \text{ earth's orbit},$$

or, in English, the Venusian orbit is 56 percent the size of the earth's orbit, or about 52 million miles.

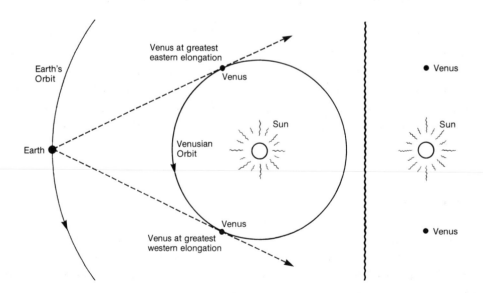

Figure 57 Looking down upon the planes of the orbits of Venus and the earth, we see that Venus never gets beyond a certain angular distance away from the sun, as seen from earth. These maximum angles are called maximum angles of elongation.

Another example: If the maximum angle east of the planet was 29 degrees, and the maximum angle west was 31 degrees, what would the actual maximum angle of elongation be? Here we have to average the east and west maximum values. The average is 30 degrees, whose sine is 0.5, making the orbital size of this fictitious planet 50 percent that of the earth's.

In a later section we will see how this simple determination of the relative orbital sizes of the planets yields a fundamental clue to the laws of planetary motion, as derived by Johannes Kepler.

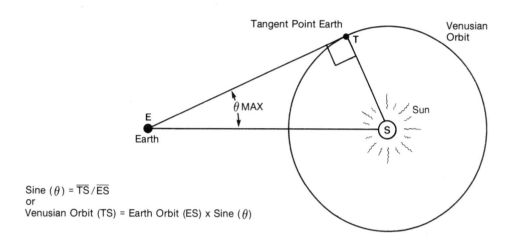

Sine (θ) = $\overline{TS}$ / $\overline{ES}$
or
Venusian Orbit (TS) = Earth Orbit (ES) x Sine (θ)

Figure 58 The right triangle relationship developed between Venus, the earth, and the sun, when Venus is at the maximum angle of elongation.

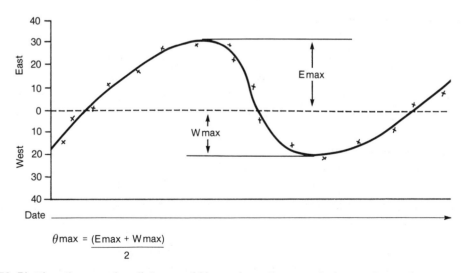

$$\theta max = \frac{(Emax + Wmax)}{2}$$

Figure 59 Plotting the angular distance of Venus from the sun during a planetarium demonstration reveals the maximum angular elongation of the planet from the sun, as seen from earth.

8   The Dynamics of Space

Tycho Brahe

Tycho Brahe (1546–1601) tested the Copernican heliocentric hypothesis using the same arguments as Aristotle. If the earth indeed moves about the sun, then since the celestial sphere was still considered to be finite, the positions of the stars should shift during the year. Tycho attempted to determine how large that shift should be.

From extended observations of the apparent positions of the stars (which, compiled into a catalog along with planetary positions, constitutes his major work), Tycho believed that the observed angular diameters of the stars was on the order of 1 minute of arc (there are 60 minutes of arc in a degree), which was actually lower than the limit of accuracy of a single observation with his astronomical equipment.

The sun's diameter, as observed in the sky, is 30 minutes of arc. Assuming that all stars have the same intrinsic linear diameters, Tycho calculated that the stars must be 30 times farther away than the sun. Tycho then predicted what the shift of stars due to the motion of the earth should be. He didn't find it from his star catalog data, hence he concluded that Copernicus's moving earth was wrong. In theory, Tycho's methods were correct, as were Aristotle's. Only two assumptions led him astray: the distances to the stars and the assumption that they were all the same size.

Actually, the stars *do* shift a bit in position during the course of the year due to the earth's motion, but the stars are more than 300,000 times farther away than the sun from the earth (for the closest one only) so the shift is extremely small; it was not detected until the mid-19th century. Still, Tycho's reasoning was correct. Figure 60 shows the geometry behind it.

As the earth moves from point 1 to point 2 in figure 60 the angular distance between any two stars in the directon of motion of the earth should increase and a relative shift should therefore occur. The earth would be moving closer to these two stars as it moves in its orbit. Tycho

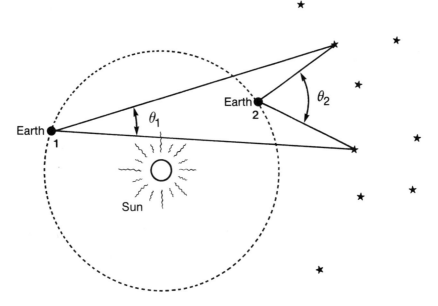

Figure 60 Angle θ_2 would be greater than angle θ_1 if the earth indeed moved around the sun. Tycho, to the limit of his observational accuracy, found that the two angles were equal, and so concluded that the earth did not orbit the sun.

found no shift because he couldn't measure stellar positions accurately enough. Today, the shifts of nearby stellar positions due to the earth's orbital motion are determined with respect to very distant stars, unlike the model in figure 60.

Tycho's observation of the path of the comet of 1577 also caused him to reject the Copernican hypothesis of a heliocentric universe. Tycho found that this comet crossed the orbits of several of the planets and therefore their orbits could not be made of solid material—the crystalline spheres—as Copernicus had believed. This removed one of Copernicus' justifications for the heliocentric hypothesis—to make the orbits of the various planets non-intersecting. Tycho was thus perfectly free to revert to a geocentric system.

Johannes Kepler

Tycho's catalog of the positions of the stars and planets was put to very good use by his assistant, Johannes Kepler (1571–1630). Kepler was an ardent believer in the Copernican system but was not satisfied with all of its points. The system still required circular orbits and a few epicycles, and it didn't do a significantly better job of predicting planetary positions than the old geocentric system. So Kepler set out to determine a better model, based upon the Copernican system, using the data collected by Tycho.

Kepler studied the orbit of Mars for several years. His goal was to create a model that could be used to calculate the positions of the planet in the future as closely as its position could be observed (statistically, about one minute of arc). Kepler tried all types of orbital paths: eccentric circles, oblate and then prolate circles, and then tried a family of projected circles called ellipses. One ellipse worked very well and so Kepler adopted what can be considered to be his **first law of planetary motion:**

Planets travel in ellipses with the sun at one focus.

What is an ellipse? Let's draw one. Stick two nails in cardboard about one inch apart. Now tie a length of string together to make a loop two inches long. Drop the loop around the nails and draw the loop taut with a pencil. Keeping the loop taut, move the pencil around the two nails. The curve that is thus drawn looks like a squashed circle. It is an ellipse, as seen in figure 61.

In figure 61, F_1 and F_2 are the foci of the ellipse. If this is a planetary orbit, the sun would be at one of the foci, say F_2. Various properties of the orbit are of interest. The eccentricity of the orbit (a measure of the degree of noncircularity) is defined as:

$$\text{Eccentricity} = CF_2/CP$$

If the sun is at F_2, then when the planet is at P, it is closest to the sun and is at its **perihelion**

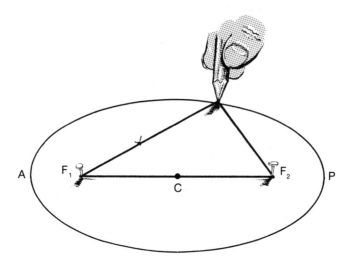

Figure 61 Construction of an ellipse.

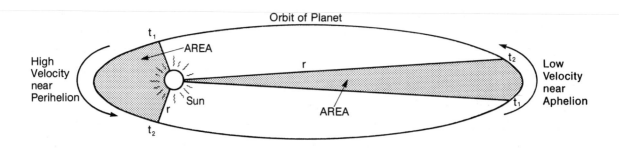

Figure 62 Illustration of Kepler's second law, his law of equal areas.

point in its orbit. When the planet is at A, it is at its farthest distance, or **aphelion** distance.

Another observation Kepler made from Tycho's observational data was that the planets didn't travel at constant rates in their orbits. When they were close to aphelion they traveled quite slowly, but they sped up as they approached the sun and attained a maximum velocity at perihelion. Kepler embodied this characteristic into his **law of equal areas.**

The radius vectors of planets (a line drawn between the planet and the sun) sweep out equal areas of space in equal intervals of time.

As the size of the radius vector diminishes, the rate of motion of the planet increases to keep the area swept out in space per unit time constant. Figure 62 is an exaggerated example. The shaded areas represent the areas swept out in time interval t_2-t_1. If the orbit was circular, the body should travel at a constant rate because the length of the radius vector is constant. In an ellipse, however, the radius vector varies in size, hence the planet's velocity varies *inversely* with the variation of the length of the radius vector.

These first two laws appeared in 1609. Kepler's search for a third unifying law, however, required a bit more work. First, he needed the sidereal periods of the planets and their semi-major axes of revolution about the sun. He derived these in a manner reminiscent of Copernicus. Kepler found, roughly:

	a(astronomical units)	P(years)
Mercury	0.38	0.24
Venus	0.72	0.62
Earth	1.00	1.00
Mars	1.52	1.88
Jupiter	5.22	11.86
Saturn	9.17	29.46

Kepler believed that there had to be a relationship between these orbital sizes and periods. In 1618 he found one and in 1619 he published it as his **harmonic law:**

The squares of the ratios of the periods of revolution of the planets are proportional to the cubes of the ratios of their mean distances from the sun . . . or

$$(P_1/P_2)^2 \ = \ (a_1/a_2)^3$$

In modern terms, with periods in sidereal years and orbital sizes in **astronomical units** (earth-sun distance equals one astronomical unit, or 1 a.u.), the equation simplifies to:

$$P^2 \ = \ a^3$$

for any body in the solar system orbiting the sun.

Example: What would the value of a be for a planet whose period around the sun is six years?

$$P^2 \ = \ 6^2 \ = \ 36 \ = \ a^3$$
$$a \ = \ 3.3 \ \text{a.u.s}$$

If the period in the example is eight years: $P^2 = 64$ and $a \ = \ 4$ a.u.

Questions

1. Take from the planetary chart several values of a (the planet's distance from the sun), cube them, and show them to be equal to the square of their periods. How well do they agree? Can you suggest a reason why the agreement is not exact?

2. What would be a for a planet whose P is 250 years? What planet has these approximate orbital characteristics.

3. Can a planet in orbit around the sun have a period of five years and a semi-major axis a of five astronomical units? If not, why not?

4. If a new inferior planet were discovered with a maximum angle of elongation of 60 degrees from the sun, what would its orbital radius be, and its sidereal period?

Galileo Galilei

Galileo Galilei (1564–1642) was Kepler's approximate contemporary in time, but not in spirit. While Kepler was the relative outsider, who led a life torn by unhappy circumstances such as the Thirty Years' War, Galileo was at the very center of the intellectual life of Italy. Galileo's work and publications were also far more straightforward than were Kepler's. Galileo had, indeed, more of a modern mind than Kepler.

Galileo's lifework can be divided into two areas. First, his application of the recently invented telescope to the observation of celestial phenomena provided strong circumstantial support for the Copernican hypothesis; and second, his development of a new system of kinematical motion went against Aristotelean thinking and laid the foundation for all subsequent concepts of motion in the Newtonian era.

Galileo's Telescopic Observations

The observations Galileo made with his telescope will only be summarized. Although he did not invent the telescope, he was the first to turn one to the heavens and tell the world about it. With a small telescope of barely one-inch aperture he discovered:

1. Mountains and valleys on the moon. This discovery indicated that the moon was a terrestrial world like the earth, which countered the Aristotelean concept of the ideal nature of the heavenly bodies.

2. Sunspots on the sun and the rotation of the sun. As with the moon, the sun became a physical object, subject to physical laws.

3. The four satellites of Jupiter, and the fact that they orbited Jupiter, which was now a

center of motion in the universe other than the earth.

4. Venus exhibits phases much like the moon. In addition, the phases were correlated with Venus's position in the sky relative to the sun and with its apparent angular diameter, both interpreted as evidence of Venus's motion around the sun.

5. The Milky Way was not a continuous band of light but was a vast realm of individual stars too faint to be seen individually by the unaided eye.

These five basic observations all aided the Copernican view of the solar system: that the system was not earth centered but centered on the sun. Galileo's telescopic observations, while important, were not as far reaching as his new system of mechanics, which discussed, in terms of observations, how objects moved on earth in response to gravity.

Galileo's Mechanics

Aristotle had decided that the nature of motion was somehow determined by the nature of the body undergoing that motion. Heavy objects had a greater tendency to fall than light objects, and for a body to remain in motion, some external force had to be continually applied. Thus, to Aristotle, the following two laws were self-evident:

1. Heavier bodies fall faster than lighter bodies.

2. If no external force acts on a moving body, then the motion of that body will diminish and the body will eventually come to rest.

The first of these two Aristotelean statements is a bit difficult to believe today, but it is consistent with his concept of *why* objects fall. If falling is a property of the body, or of the body's mass, then the rate of fall should be a function of that mass.

In a story often told, Galileo decided to test this idea. He therefore conducted (or directed someone else to carry out) a test, having an assistant climb to the top of the Tower of Pisa, drop two objects of unequal weight, and then observe how they fell. As the story goes, the objects fell at the same rate and reached the ground at the same time. From this, Galileo concluded that the rate of fall was independent of the object's mass (or weight, for the two properties of the object were not separated at the time) and was somehow a function only of an external force acting on the two bodies equally. Some historians wonder if this experiment ever actually took place, but true or not, it clearly illustrates the need for direct tests of assumptions.

Galileo tested the second Aristotelean statement by experimenting with inclined planes. He found that, as in figure 63, a ball rolling down an inclined plane (in the direction of gravity) is seen to speed up, and one rolling up an inclined plane (opposite the direction of gravity) will slow down. The average case between the two comes, Galileo reasoned, when the inclined plane is horizontal (no forces acting, assuming also there is no friction). In the absence of external forces, no acceleration or deceleration will occur. If there was an original motion to the object, that motion will remain, and the speed of the ball will be constant. While Galileo

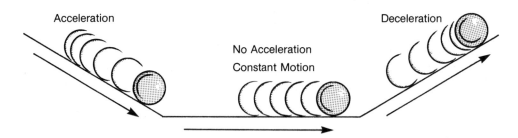

Figure 63 Inclined planes

did experiment with inclined planes, the logical sequence just described probably was derived from a combination of many different types of observations and some creative speculation.

To Galileo, the natural state of motion, in the absence of acceleration or deceleration, was *constant* motion, *not* rest, as Aristotle had stated. Why then do objects slow down once in motion, as we often see in the real world? The real world contains all sorts of external forces, such as rolling friction, air friction, and sliding friction. These external forces act on moving bodies in a direction opposing their motion, and bring them to "rest."

Galileo did not attempt to relate his work directly to celestial motion. If he had done so, he would have had to modify a few things. First, celestial objects in orbits do *not* exhibit constant motion. Motion in an orbit is accelerated motion. Even if the speed of a planet were constant, its constantly changing direction would have to be accounted for by the action of an external force. Newton was later able to show that an external force was indeed acting to keep the planets in their orbits. This was his law of universal gravitation. To be able to understand the basis of this, we have to complete our discussion of Galileo's work.

Galileo found from his inclined plane studies that the rate of fall of an object was independent of the mass or size of the object. In fact, the distance an object would fall, all the while undergoing a constant acceleration due to the gravitational field of the earth (a property Galileo did not specify), was a function of the time of fall alone and was independent of everything else. What he found can be neatly summarized in one important equation:

S (Distance of fall) = ½ × a (acceleration of fall) × t²(time of fall squared)

or

$$S = ½ × a × t^2$$

If the distance of fall was constant, then the time of fall would always be constant as long as no other forces were acting *in the direction or opposite the direction of fall*. If other forces were acting perpendicular to the direction of fall (in this case, perfectly horizontal forces), they would not alter the time of fall. To illustrate the situation, we suggest the following thought experiment:

Aim a gun (sling shot, arrow, marble rolling off a table, tennis ball, etc.) horizontally and fire parallel to the surface of the earth. At each successive firing, pour more gunpowder (or apply more motive force) so that the muzzle velocity is greater each time. Note how long it takes each bullet to fall to the ground, and where each bullet falls. Result: All bullets will fall to the ground in the same amount of time. The only difference will be that bullets fired with higher muzzle velocity will travel farther down the range, as is seen in figure 64.

From figure 64, the only difference in each firing is that the curvature of the paths of the low- and high-velocity bullets will be different. The higher the muzzle velocity, the less will be the curvature of the path of the bullet toward the ground. If the earth were flat, then all bullets fired would fall to the ground at the same time. But the earth is not flat, and this is where Newton comes in.

Isaac Newton

Newton decided, possibly while sitting under an apple tree (again historians wonder about this), that the phenomenon of how bodies fall on earth could be extended to describe how the moon moves around the earth and how the planets move around the sun. If the moon was simply "falling" around the earth, then its rate of fall should only be a function of the acceleration of fall (due to the force of attraction caused by the earth). In order to think this through, Newton modified Galileo's scene in figure 64.

As figure 65 demonstrates, at a particular muzzle velocity, the ratio of the rate of fall to the downrange distance traveled (the curvature of the path) will equal the curvature of the earth, and so the distance of the bullet from the ground will never change. If there is no air resistance impeding the motion of the bullet after it leaves the gun, the bullet will continue to "fall" around the earth indefinitely. The only initial force was from the gun. The constantly acting force is the earth's gravity.

Newton generalized this argument to discuss how the moon fell around the earth. He was able to show that if the moon had an initial motion tangential to the earth (parallel to the surface of the earth) then the only force required to keep the moon in its orbit was directed toward the earth itself and was a manifestation of the gravitational attraction of the earth.

To show this, Newton found that the acceleration of the moon falling toward the earth was 1/3600 of the acceleration experienced by an object falling at the surface of the earth.

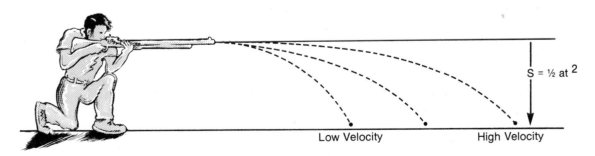

Figure 64 The time of fall of an object fired horizontally is independent of its muzzle velocity and depends only on the distance of fall.

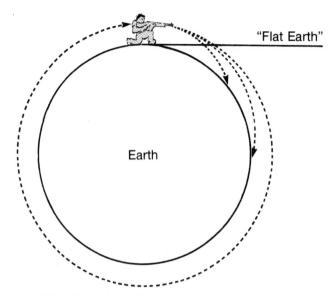

Figure 65 Objects have to fall "farther" to reach ground on a curved earth. If the curvature of the path equals the curvature of the earth, the object will never reach the ground.

Since the moon is 60 times farther from the center of the earth than is a body (an apple) at the earth's surface, the force of attraction due to the earth's gravity not only diminishes as one moves away from earth but it varies in a special way. As the distance of any object increases from the center of the earth, the force of gravitational attraction on that object due to the earth decreases by the square of the distance. For example, if the distance of an object were doubled, the force of attraction would be cut by four times.

Without the earth's attractive force, the moon would move in a straight line at constant velocity (as would be deduced from Galileo's work). With the earth's attraction, however, the moon's path is curved toward the earth (see figure 66). The amount of curvature (the amount of fall per unit time) is just enough to keep the moon close to the same distance from the earth, moving in an elliptical path. Newton was able to determine what this rate of fall was, and from it was able to find that the force of attraction of the earth was inversely proportional to the square of the distance from the center of the earth.

Newton was able to say all this only after he generalized Galileo's work on mechanics and motion, and expressed them as three laws of motion:

1. An object in motion will remain in motion unless acted upon by an outside force. Thus, if all forces = 0, there will be no acceleration, and so velocity and momentum remain constant, or:

Mass × Velocity = Constant.

2. If an outside force acts upon that object, the motion of that object will change in proportion to the implied force, and inversely as the mass of the object:

$$\text{Force} = \text{Mass} \times \text{Acceleration}$$
$$= \text{Mass} \times (V_2 - V_1)$$

3. For every force implied on an object, there is an equal but opposite force applied by the object on the agent producing the force. (Law of action and reaction)

From the second law, we were finally able to distinguish between mass and weight. Our "weight" on earth is really a manifestation of the fact that we are subject to the earth's gravitational force field, and hence the mass that comprises our bodies experiences a force of attraction by the earth.

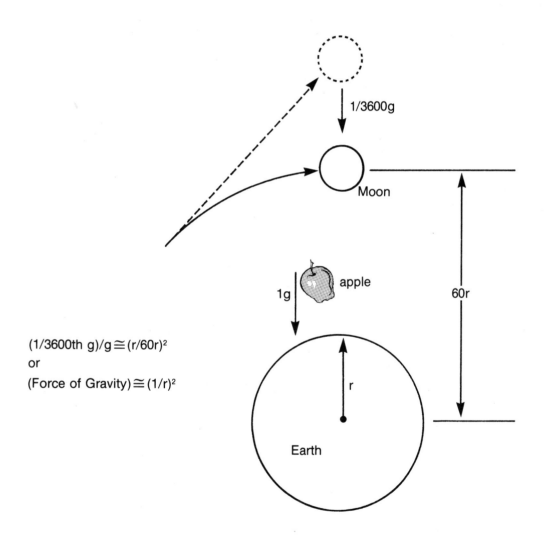

$(1/3600\text{th } g)/g \cong (r/60r)^2$
or
$(\text{Force of Gravity}) \cong (1/r)^2$

Figure 66 An apple falling near the surface of the earth experiences 1g acceleration caused by the earth. The moon experiences 1/3600th g acceleration caused by the earth because it is 60 times farther away from the earth's center than is the apple.

From these basic laws of motion, Newton was able to generalize his derivation of the inverse square law of gravity:

$$F = G \times (m_1 \times m_2)/d^2$$

where F is the force of attraction between any two bodies in the universe, G is a constant of proportionality (the gravitational constant), m_1 is the mass of one object and m_2 the mass of the other object, and d is the distance between the two objects. This is Newton's universal law, which describes much of the dynamic behavior of celestial objects, from the motion of the moon and earth around the sun to the formation of stars and the motions of stars in binary systems.

With these laws, Newton was able to go back to Kepler's three laws of planetary motion and rederive them mathematically to show that his universal law of gravity could explain the motions of planets. Newton's work, therefore, was the synthesis of the mechanics of Galileo and the studies of planetary motion of Kepler.

One of the most important equations to come out of Newton's synthesis was his generalization of Kepler's third law, $P^2 = a^3$. Newton showed that the masses of the objects in orbit could be included in the equation, along with the universal constant of gravity and the constant pi = 3.14:

$$P^2 = 4 \times pi^2 \times a^3/(G \times (m_1 + m_2))$$

Questions

1. If two satellites are circling the earth, and one is twice as far from the earth as the other, what is the ratio of the earth's force of attraction on the two satellites?

2. If Galileo found that Jupiter's satellites had orbital sizes in the ratio of 1:2:4:8, and periods in the ratio of 1:4:16:64, would that have been a verification of Kepler's third law? If not, why not?

3. If the period of the moon was not 27 days but 54 days around the earth, and its distance unchanged, what would be the resultant combined mass of the earth-moon system?

Hint: The most efficient way to answer is to place the equation into a form of a ratio taking what is presently accepted, and comparing that to the new condition.

Looking at Question 3, with less mass, the force of attraction would be less, and the response to that force (acceleration) is less. With less acceleration it would take the moon longer to travel around the earth.

This brings up another facet of the problem: In any gravitational system, both bodies move around their mutual center of mass. The earth is not the stationary center of the moon's orbit. Both the earth and the moon travel about one another. The "stationary" center of mass is closer to the earth because the earth is more massive, about 81 times more massive than the moon. The center of motion is therefore 81 times closer to the earth than to the moon. It is about 3,000 miles from the center of the earth, or within the earth's surface!

4. If the periods of planets around the sun are measured in years and their orbital sizes in a.u.s, then the combined mass of the sun and planet can be determined in terms of "solar masses" (one solar mass $= 2 \times 10^{33}$ grams) by Kepler's equation modified by Newton:

$$M_1 + M_2 = a^3/P^2$$

If the earth traveled around the sun in half a year, but its distance remained the same, what then would be the mass of the earth and sun combined?

5. The same equation can be applied to determine the masses of stars in double-star systems. If an astronomer determines that a double-star system has a mutual period of revolution of 10 years, and a semi-major axis distance (the size of the orbit between the two components) of 10 a.u.s, what is the combined mass of the two stars in the system?

Appendix   Making a Simple Astrolabe

Historically, astrolabes refer to a range of measuring and calculating devices that developed more than one thousand years ago to measure time and to calculate place on earth. Astrolabes may well have originated in Greek times, but the earliest examples come to us from the 9th through the 11th centuries, and were constructed by Arab and Persian craftsmen. Astrolabes from antiquity were able to measure the altitudes of stars, and, given the date, were used to calculate from that one observation the time of night. The astrolabe we will construct will measure altitudes only.

Cut the template provided here out of the page (or make a copy) and paste it onto a piece of cardboard. Then tape a straw to the 90-degree side of the template. Finally, tie a one-foot string with a small weight attached to the center of the template protractor.

By sighting through the straw at any object, its altitude above the physical horizon can be measured directly from where the string crosses the protractor.

Since we will be looking for the altitude of the sun at noon in this experiment, the straw will become a sighting tube used in the following manner: 1) Do not sight on the sun with your eye directly! 2) Hold the astrolabe at a right angle to the sun and place your hand behind the straw so that you will see the shadow of the straw on your hand. When the straw is pointed directly at the sun, you will see a small circle of light on your hand. The astrolabe is now pointing directly at the sun and you can read the altitude of the sun directly from where the string crosses the protractor.

Astrolabe Template

Once a week at local apparent time noon, when the sun is crossing your observer's meridian, measure the altitude of the sun. Over the course of several weeks, you will see that the noontime altitude of the sun changes. Plotting over an entire season (crossing an equinox and solstice date), you will be able to determine your latitude and the obliquity of the ecliptic. While this takes months in the real sky, it can be done more quickly in the planetarium.

Template:

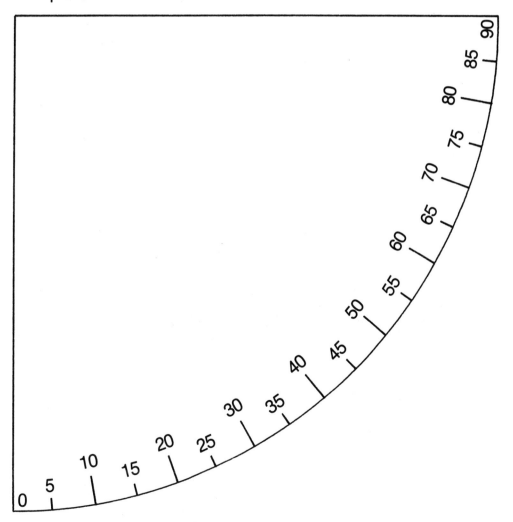

Glossary

alternate interior angles. Angles formed when a straight line (the transversal) intersects two parallel lines. Angles on opposite sides of the transversal are equal.

altitude. 1) The linear height of an object above the earth's surface. 2) The angular distance of an object on the celestial sphere from the horizon.

aphelion. The point on a closed orbit around the sun at greatest distance from the sun.

astrolabe. A device to measure time and calculate place on earth. In the form used in this book, it is also known as a quadrant.

astronomical unit. The mean distance between the earth and sun. A standard unit of astronomical distance.

autumnal equinox. Point on the celestial equator where the sun passes moving south. Occurs around September 23.

cardinal points. The four major compass points north, south, east and west, on the physical horizon.

celestial equator. The extension of the earth's equator onto the celestial sphere.

celestial meridian. The extension of an observer's meridian of longitude out onto the celestial sphere.

celestial sphere. The apparent plane of the sky.

circumpolar. A celestial object that is always above the horizon from a given point on earth.

cosine. The ratio of the adjacent side of a right triangle to its hypotenuse

declination. The angular distance of a celestial object from the celestial equator.

deferent. A large rotating circle used to represent the primary motion of celestial bodies around the earth in the Greek geocentric universe.

diurnal motion. The apparent daily motions of objects on the celestial sphere, as seen from a moving earth.

ecliptic. The apparent path of the sun among the stars, caused by the revolution of the earth around the sun.

epicycle. A small rotating circle whose center is on a larger rotating circle.

equation of time. The difference between standard time and Local Apparent Time, caused by the uneven motion of the apparent sun on the ecliptic through the course of the year.

equator. A circle on a sphere equidistant between its poles. The earth's equator is equally distant from its north and south poles.

equinox. 1) Equal lengths of daylight and night. 2) Two dates of the year (usually March 21 and September 23) when the sun is on the celestial equator.

geocentric. Centered upon the earth.

globular cluster. A spherical cluster of up to one million stars.

great circle. The largest circle that can be inscribed on a sphere. The diameter of a great circle is equal to the diameter of the sphere it is inscribed upon.

heliocentric. Centered upon the sun.

horizon. The apparent or actual line separating the visible sky from the earth. The *physical horizon* is a plane perpendicular to the direction of gravity.

hour angle. The angular distance of an object from the observer's meridian, measured at the point of intersection of the hour circle passing through the object and the celestial equator.

hour circle. A circle on the celestial sphere passing through the north and south celestial poles.

hypotenuse. The side of a right triangle opposite the right angle.

inferior. An orbit smaller than a reference orbit (usually the earth's orbit).

latitude. The angular distance of a point on earth north or south of the equator.

Local Apparent Time. Time as measured by the passage of the sun in the sky.

longitude. The angular distance, east or west, of a point on the surface of the earth from the Greenwich Meridian, measured in units of time or angle (see meridian of longitude).

meridian of longitude. A great circle passing through the north and south poles of the earth and intersecting the equator at a specific longitude point.

Milky Way. That portion of our home galaxy that is visible from earth.

North Celestial Pole. A point on the celestial sphere directly above the North Pole of the earth.

observer's meridian. The specific meridian of longitude of an observer, extended onto the celestial sphere.

perihelion. The point on a closed orbit around the sun nearest the sun.

retrograde. Reverse or backwards motions on the celestial sphere; westward motion.

revolution. The orbital motion of a celestial body.

right ascension. The celestial longitude of an object, measured eastward from the vernal equinox along the celestial equator to the point where the object's hour circle intersects the equator.

sidereal time. Time as determined by the motions of stars.

sine. The ratio of the opposite side of a right triangle to its hypotenuse.

solstices. Points on the ecliptic where the sun is at greatest angular distance north (summer) and south (winter) of the celestial equator.

solar time. Time as determined by the apparent movement of the sun. A measure of Local Apparent Time.

South Celestial Pole. A point on the celestial sphere directly above the South Pole of the earth.

standard meridian. One of 24 meridians of longitude spaced 15 degrees (or one hour apart) around the earth from which standard time is measured.

superior. An orbit larger than a reference orbit (usually the earth's orbit).

tangent. 1) The ratio of the opposite side of a right triangle to the adjacent side. 2) A line intersecting a circle or curve at one point.

vernal equinox. The intersection of the celestial equator and the ecliptic where the sun is moving from the southern into the northern hemisphere. Occurs on March 21.

zenith. The vertical point on the celestial sphere, 90 degrees from the physical horizon.

Selected Bibliography

The following textbooks, popular reviews and historical works are useful in gaining a deeper understanding of the topics discussed in these lecture notes.

Textbooks

Abell, George. *Exploration of the Universe*. 3d ed. New York: Holt, Rinehart and Winston, 1975.

Numerous versions and editions available. Still considered one of the standard testbooks on general astronomy and one of the most reliable.

Pasachoff, Jay. *Astronomy: From the Earth to the Universe*. Philadelphia: Saunders, 1983.

Numerous versions and additions exist of this popular elementary text, under similar titles.

Rogers, Eric. *Astronomy for the Inquiring Mind*. Princeton: Princeton University Press, 1982.

Well illustrated introduction to the history of the astronomical topics treated here. Good exposition of Greek geometry applied to astronomy.

Smart, William. *Text-Book on Spherical Astronomy*. Cambridge: Cambridge University Press, 1962.

Excellent mid-level technical review available in many paperback editions. First published in 1931.

Popular Reviews

Allen, Richard Hinckley. *Star Names*. New York: Dover, 1963.

Excellent source for the origins of the names of stars and constellations. Provides detailed descriptions of objects within each visible constellation.

Fuller, E., ed. *Bulfinch's Mythology*. New York: Dell, 1967.

One of many modern versions of Thomas Bulfinch's 19th-century classic treatment of mythology that includes stories of the constellations from Greece ("The Age of Fable").

Gamow, George. *Gravity*. New York: Doubleday/Anchor, 1962.

A lucid elementary introduction to the theory of gravitation.

Hamilton, Edith. *Mythology*. New York: Mentor/New American Library, 1962.

Originally published in 1940. An excellent introduction to classical mythology.

Hoyle, Fred. *Astronomy*. New York: Doubleday, 1962.

A popular treatment that contains good descriptive material on planetary motion, Newtonian celestial mechanics, and the development of the theory of gravitation.

Norton's Star Atlas. Cambridge, Mass.: Sky Publishing Corporation, 1973.

Sixteenth edition of a famous bound series of maps containing all stars to the visual limit, with non-stellar objects identified and described. First published in 1910.

Rey, H. A. *The Stars, A New Way to See Them*. Boston: Houghton Mifflin, 1962.

An ingenious set of star maps with modern simplified figures, together with a good introduction to visible sky astronomy. Numerous editions have been published.

Historical Works

Those cited are generally available in paperback or in most libraries.

Cohn, I. Bernard. *The Birth of a New Physics*. New York: Doubleday/Anchor, 1960.

Drake, Stillman. *Discoveries and Opinions of Galileo*. New York: Doubleday/Anchor, 1957.

Dreyer, J. L. E. *History of Astronomy from Thales to Kepler*. 2d ed. New York: Dover, 1953.

Numerous reprints exist of this classic history.

Hall, A. Rupert. *From Galileo to Newton: 1630–1720*. London: Fontana, 1970.

Koestler, Arthur. *The Sleepwalkers*. New York: Grosset and Dunlap, 1970.

Contains an excellent chapter on Kepler entitled *The Watershed*, which has been published separately.

Kuhn, Thomas. *The Copernican Revolution*. Cambridge, Mass.: Harvard University Press, 1957. (Paperback edition available from Random House.)

Van Helden, Albert. *Measuring the Universe*. Chicago: University of Chicago Press, 1985.

Detailed midlevel historical review of methods used to determine the distances and sizes of celestial objects, from antiquity to Newton and Edmond Halley.

Answers to Questions

From page 24:

1. No. Maximum altitude will be only about 75 degrees. It will be seen at the zenith at points between 23½ degrees north and 23½ degrees south latitudes.

2. 45 degrees; 20 degrees; − 20 degrees.

3. 90 degrees—always the largest declination and it is always circumpolar.

4. The least declination that is circumpolar from Washington (latitude 38 degrees) is 90–38 degrees or 52 degrees.

5. Range of declination that will be circumpolar = 90 − latitude to 90 degrees.

6. 45 degrees.

7. Measuring the altitude of the North Star (Polaris), roughly, or the meridian altitude of any star of known declination, and making the appropriate corrections.

From page 32:

1.
   a. 75 degrees west.
   b. EDST = 12:00 = 11:00 a.m. EST, so you are 15 degrees east of 75 degrees longitude, or at 60 degrees west longitude.
   c. 20 minutes ahead, or at 70 degrees west longitude.
   d. 16 minutes behind, or at 79 degrees west longitude.

2.
   a. 45 degrees west.
   b. 90 degrees west.
   c. 30 degrees west.

From page 37:

1. R.A. of Orion = 6 hours
   Sun is at 18 hours R.A.
   So season is early winter (late December near the solstice).

2. Sun is at 6 hours R.A., so season is summer.

3. In the evening sky. It will transit two hours after the sun, or at about 2:00 p.m.

4. Sun is at 18 hours R.A. on December 21. Therefore on November 21, one month earlier, the sun's R.A. is two hours less, or at 16 hours. The sidereal time will be 16 hours at L.A.T. noon.

From page 43:

1. On May 15, the sun's declination is between 19 and 20 degrees. If it is at 20 degrees, and its meridian altitude is 45 degrees, the A.C.E. will be 45 − 20 = 25 degrees. Thus:

$$90 - \text{L.A.T.} = \text{A.C.E.} = 25$$
$$\text{L.A.T.} = 90 - 25 = 65 \text{ degrees north}$$

   Your longitude is four minutes west of 75 degrees. At four minutes per degree of longitude, you are therefore at 76 degrees W longitude.

2. No. Maximum northerly declination of the sun is 23½ degrees. At the latitude of Washington (38 degrees), the sun will not obtain a meridian altitude of greater than 75 degrees.

3. Latitude = 23½ degrees north. The Tropic of Cancer.

4. On the equinoxes: on or around March 21 and September 23.

5. Solar declination = 23½ degrees.
   Latitude = − 38 degrees.
   A.C.E. = 52 degrees *south*.
   Altitude of the sun = 52 − 23½ = 28½ degrees.

From page 45:

1. March 21.

2. March 21.

3. Approximately November through mid-February:
   At latitude 75 degrees N, A.C.E. = 90 − 75 = 15 degrees, so sun's declination has to be between − 15 and − 23½ degrees.

4. 66½ degrees north.

From page 53:

1. Pollux rises about four hours after sun, or at 10:00 a.m. It crosses the meridian at S.T. = 7½ hours, when H.A. of sun is 5½ hours, or at 5:30 p.m. It will set at about 2:00 a.m.

2. Capella (declination = 46 degrees) goes through the zenith from latitude = 46 degrees north. On May 9 it will pass the meridian at 2:20 p.m. EST, as seen from longitude 70 degrees west.

3. Latitude: 63.5 degrees north.
   Longitude: 112 degrees west.

From page 59:

1. 10:30 p.m.

2. (A) winter: 75 degrees.
   (B) summer: 28 degrees.
   In general, the winter moon is more effective in illuminating the night landscape because it rises higher in the sky.

3. First-quarter moon has to be on the celestial equator; this will occur when the sun is at either solstice or around June 21 or December 21.

4. Around 8:00 p.m.

5. Around 3 a.m.

6. Around December 21.

From page 64:

1. At least 12, sometimes 13. No, only from narrow positions of the earth's surface bathed in the lunar shadow. Lunar eclipses would be visible on the night side of the earth.

2. Lunar eclipses can be seen by anyone who can see the moon at the time of eclipse. Solar eclipse visibility is confined to the narrow eclipse path defined by the width of the moon's shadow on the earth's surface, never more than several dozen to one or two hundred miles wide.

3. *Connecticut Yankee in King Arthur's Court.* Depicted as a fearsome and unpredictable happening, and anyone with the ability to predict such things had powers that were respected and feared.

4. Far more solar and lunar eclipses would occur. The angular diameter of the moon would be twice what it is now, and the size of the shadow of the earth at the point where the moon crosses it would be larger. Also, the period of the lunar orbit would be less by just under a factor of 3, so eclipses might occur more frequently. Although the size of the lunar shadow on the earth's surface would be larger, the moon's angular orbital speed would be faster still, so the duration of eclipses might actually be less.

From page 86:

1.

Mercury:	a = 0.38		a^3 = 0.0548
	P = 0.24		P^2 = 0.0576
Venus:	a = 0.72		a^3 = 0.373
	P = 0.62		P^2 = 0.384
Mars:	a = 1.52		a^3 = 3.51
	P = 1.88		P^3 = 3.53
Jupiter:	a = 5.22		a^3 = 142.24
	P = 11.86		P^2 = 140.66
Saturn:	a = 9.17		a^3 = 771.10
	P = 29.46		P^2 = 867.90

The lack of perfect agreement is due largely to Kepler's inexact knowledge of the orbital sizes of the planets and to his complete lack of knowledge of the considerable masses of the Jovian planets, which enter into Newton's modifications of Kepler's third law.

2. $P^2 = a^3 = (250)^2 = 62,500$

   $a = 39.7$ a.u., approximately the size of Pluto's orbit.

3. No. Does not obey Kepler's third law: $a^3 = 125$

   $P^2 = 25$

4. Sine $60° = 0.866$

   $a = 0.866$

   $a^3 = 0.649 = P^2$

   $P = 0.81$ a.u.

From page 92:

1. $F_1/F_2 = (R_2/R_1)^2 = (2/1)^2 = 4$

2. No verification; the squares of the ratios of the periods of the satellites do not agree with the cubes of the ratios of their orbital sizes.

3. $(P_1/P_2)^2 = (m_1 + m_2)_2/(m_1 + m_2)_1 = (27/54)^2 = (1/2)^2 = 1/4$

   So the new mass of the system would be one-fourth the mass of the original system.

4. Four times the present mass, or four solar masses.

5. 10 solar masses.

Table of Sines

ANGLE	SINE	ANGLE	SINE
18	.31	20	.34
22	.37	24	.40
26	.44	28	.47
30	.50	32	.53
34	.56	36	.59
38	.62	40	.64
42	.67	44	.69
46	.72	48	.74
50	.77	52	.79
54	.81	56	.83
58	.85	60	.87

Sample Data Sheet for Seasons Exercise

Latitude: _____

	SUNRISE (S.T.)	(AZ)	NOON (S.T.)	(ALT)	SUNSET (S.T.)	(AZ)	#HRS LIGHT
Date							
March 21							
June 22							
Sept. 23							
Dec. 22							

Latitude: _____

Date							
March 21							
June 22							
Sept. 23							
Dec. 22							

Data Sheet: Telling Time by the Phases of the Moon

SEASON		R.A. SUN	RISE S.T.	RISE L.A.T.	MER. PASS S.T.	MER. PASS L.A.T.	SET S.T.	SET L.A.T.
SPRING	NEW							
	1ST							
	FULL							
	3RD							
SUMMER	NEW							
	1ST							
	FULL							
	3RD							
FALL	NEW							
	1ST							
	FULL							
	3RD							
WINTER	NEW							
	1ST							
	FULL							
	3RD							

USE: L.A.T. = H.A.(Sun) + 12 hrs. *and* H.A.(Sun) = S.T. − R.A.(Sun)

Orbit of Venus Data Sheet

DATE	Venus (E/W)	Angular Distance

Index